Springer Theses

Recognizing Outstanding Ph.D. Research

Aims and Scope

The series "Springer Theses" brings together a selection of the very best Ph.D. theses from around the world and across the physical sciences. Nominated and endorsed by two recognized specialists, each published volume has been selected for its scientific excellence and the high impact of its contents for the pertinent field of research. For greater accessibility to non-specialists, the published versions include an extended introduction, as well as a foreword by the student's supervisor explaining the special relevance of the work for the field. As a whole, the series will provide a valuable resource both for newcomers to the research fields described, and for other scientists seeking detailed background information on special questions. Finally, it provides an accredited documentation of the valuable contributions made by today's younger generation of scientists.

Theses are accepted into the series by invited nomination only and must fulfill all of the following criteria

- They must be written in good English.
- The topic should fall within the confines of Chemistry, Physics, Earth Sciences, Engineering and related interdisciplinary fields such as Materials, Nanoscience, Chemical Engineering, Complex Systems and Biophysics.
- The work reported in the thesis must represent a significant scientific advance.
- If the thesis includes previously published material, permission to reproduce this must be gained from the respective copyright holder.
- They must have been examined and passed during the 12 months prior to nomination.
- Each thesis should include a foreword by the supervisor outlining the significance of its content.
- The theses should have a clearly defined structure including an introduction accessible to scientists not expert in that particular field.

More information about this series at http://www.springer.com/series/8790

Abhay Shastry

Theory of Thermodynamic Measurements of Quantum Systems Far from Equilibrium

Doctoral Thesis accepted by University of Arizona, USA

Abhay Shastry
Department of Chemistry
University of Toronto
Toronto, ON, Canada

ISSN 2190-5053   ISSN 2190-5061   (electronic)
Springer Theses
ISBN 978-3-030-33576-2   ISBN 978-3-030-33574-8   (eBook)
https://doi.org/10.1007/978-3-030-33574-8

This Springer imprint is published by the registered company Springer Nature Switzerland AG.
The registered company address is: Gewerbestrasse 11, 6330 Cham, Switzerland

To my mother (Amma), who instilled in me scientific temper over dogmatism

Supervisor's Foreword

In this Springer Thesis, Abhay Shastry presents several related advances in the field of quantum thermodynamics, with a focus on the relation between thermodynamic measurement protocols and local observables in quantum systems far from equilibrium.

The central result is an ingenious proof that a joint local measurement of temperature and voltage in a system of fermions far from equilibrium exists and is unique, placing the concept of local temperature in quantum systems on a rigorous mathematical footing for the first time. As an intermediate step, a proof of the positivity of the Onsager matrix of thermoelectric linear response theory is given, a statement of the second law of thermodynamics that had lacked an independent proof for 85 years.

This thesis also poses a question related to the third law of thermodynamics: Can a local temperature arbitrarily close to absolute zero be achieved in a nonequilibrium quantum system? A systematic analysis using quantum transport theory shows that absolute zero can never be reached, but that one can come arbitrarily close.

In a more practical vein, a new scanning probe methodology to measure the local temperature of an electron system using purely electrical techniques is proposed, which could enable improving the spatial resolution of thermometry by several orders of magnitude.

Finally, a new mathematically exact definition for the local entropy of a quantum system in a nonequilibrium steady state is derived. Several different measures of the local entropy are discussed, relating to the thermodynamics of processes that a local observer with varying degrees of information about the microstates of the system could carry out.

Together, the themes investigated by Abhay Shastry in this Springer Thesis illuminate some of the key mysteries in the thermodynamics of quantum systems far from equilibrium.

Tucson, AZ, USA
staffordphysics92@gmail.com
http://www.physics.arizona.edu/~stafford/

Charles A. Stafford

Acknowledgements

First and foremost, I am very grateful to my doctoral adviser Charles A. Stafford for his constant support and patient mentorship over the years. I have gained immensely from his in-depth knowledge of theoretical physics; I learned to critically examine scientific claims and hold my research to the highest objective standards. I will certainly miss our daily lunches and scientific discussions over coffee.

I would like to thank Oliver Monti and Brian LeRoy for useful discussions. I would also like to thank Janek Wehr for his lectures in mathematical physics which helped me with parts of this work. I am thankful to the seniors in my group, Justin Bergfield and Joshua Barr, for the help I received during my initial years. It has also been my great pleasure to work with enthusiastic undergraduate students Yiheng Xu, Marco Jimenez, Sosuke Inui, and Marcus Rosales. This book is a refinement of my doctoral dissertation and also includes many additional results and discussions. I would like to thank my postdoctoral supervisor Dvira Segal for useful discussions as well as for her understanding regarding my time commitment for this book.

My parents, Sachidanand (Appa) and Kathyayini (Amma), have been supportive throughout my graduate studies and I am grateful to them. Throughout my upbringing in India, my parents strived to afford me the unique luxury of freely pursuing my passion and have supported me constantly. I am very grateful to my sister Tejaswini and my brother-in-law Karthik who have always been concerned for my well-being.

I would like to thank my colleagues Steve Steinke, Alex Abate, Adarsh Pyarelal, Pedro Espino, Souratosh Khan, Sarah Jones, Rebekah Cross, and Sophia Chen for their entertaining company and friendship. I am grateful to Mike Strangstalein for his advice over the past couple of years. Naturally, it is impossible to mention all the people who have helped me along the way given the space constraints. I thank all my many friends in the wider Tucson community who made my stay so memorable! I am also grateful to my many teachers and friends from India. I would like to thank my highschool friends K.P. Nagarjun and Abhiram Muralidhar, my undergraduate friend Varun Narasimhachar, and my undergraduate adviser Tarun Deep Saini.

I am very grateful to my partner Shraddha Satish Thumsi for her unwavering love and support over the past few years. I cannot imagine having a better partner.

Finally, I would like to thank the US Department of Energy for the financial support I received under award number DE-SC0006699. I believe the work presented here advances the scientific mission of the Department of Energy.

Toronto, ON, Canada   Abhay Shastry

Parts of This Book Have Been Published in the Following Journal Articles or Preprints

- Abhay Shastry and Charles A. Stafford. "Cold spots in quantum systems far from equilibrium: Local entropies and temperatures near absolute zero". In: Phys. Rev. B 92 (24 Dec. 2015), p. 245417. https://doi.org/10.1103/PhysRevB.92.245417. http://link.aps.org/doi/10.1103/PhysRevB.92.245417.
- Abhay Shastry and Charles A. Stafford. "Temperature and voltage measurement in quantum systems far from equilibrium". In: Phys. Rev. B 94 (15 Oct. 2016), p. 155433. https://doi.org/10.1103/PhysRevB.94.155433. http://link.aps.org/doi/10.1103/PhysRevB.94.155433.
- Charles A. Stafford and Abhay Shastry. "Local entropy of a nonequilibrium fermion system". In: The Journal of Chemical Physics 146.9 (Feb. 2017), p. 092324. https://doi.org/10.1063/1.4975810. Eprint: http://dx.doi.org/10.1063/1.4975810.
- Abhay Shastry, Sosuke Inui, and Charles A. Stafford. "Scanning tunneling thermometry". In: ArXiv e-prints 1901.09168 (Jan. 2019).
- Abhay Shastry, Yiheng Xu, and Charles A. Stafford. "The third law of thermodynamics in open quantum systems". In: The Journal of Chemical Physics 151.6 (Aug. 2019), p. 064115. Eprint: https://doi.org/10.1063/1.5100182.
- Abhay Shastry and Charles A. Stafford "Exact local entropy, entropic inequalities, and their relation to available information for a nonequilibrium fermion system". In Preparation.

Contents

List of Abbreviations

DOS	Density of states
EA	Engquist–Anderson (definition)
HOMO	Highest occupied molecular orbital
LUMO	Lowest unoccupied molecular orbital
(L)(P)DOS	(Local) (Partial) Density of states
NEGF	Nonequilibrium Green's function (formalism)
NMR	Nuclear magnetic resonance
STM	Scanning tunneling microscope (or microscopy)
STP	Scanning tunneling potentiometer (or potentiometry)
STT	Scanning tunneling thermometer (or thermometry)
SThM	Scanning thermal microscopy
WF	Wiedemann–Franz (law)

Chapter 1
Introduction

1.1 Synopsis

Thermodynamics is a well-established field which studies systems in equilibrium and provides some of the most general results in all of physics. Unluckily, the vast majority of systems encountered in nature are out of equilibrium. Thermodynamic descriptions of nonequilibrium systems are a formidable theoretical challenge and most results have been obtained under the assumption of a local equilibrium. Outside such an assumption, definitions of basic thermodynamic state variables such as temperature and voltage are muddled with a competing panoply of "operational" definitions. The work presented in this book provides a mathematically rigorous foundation for temperature and voltage measurements in quantum systems far from equilibrium. We show the existence and uniqueness of temperature and voltage measurements for any quantum fermion system in a steady state, arbitrarily far from equilibrium, and with arbitrary interactions within the quantum system. We show that the uniqueness of these measurements is intimately tied to the second law of thermodynamics. In achieving this goal, we prove the positive-definiteness of the Onsager matrix in the context of thermoelectric transport which had only been a phenomenological statement for the past 85 years. The validity of the laws of thermodynamics far from equilibrium are discussed and particular attention is paid to the second and third laws. A detailed discussion of what constitutes an ideal measurement is also included. These results have fundamental implications for the field of scanning probe microscopy. We propose a method for imaging temperature fields in nanoscopic quantum conductors where we anticipate a remarkable improvement in the spatial resolution by over two orders of magnitude. Finally, we discuss the entropy of a quantum system far from equilibrium. We obtain a hierarchy of inequalities for the entropy of the quantum system and discuss its intimate relation to the information available from a measurement. Proofs of the third law of thermodynamics are given for open quantum systems in equilibrium and

© Springer Nature Switzerland AG 2019   1
A. Shastry, *Theory of Thermodynamic Measurements of Quantum Systems Far from Equilibrium*, Springer Theses, https://doi.org/10.1007/978-3-030-33574-8_1

in nonequilibrium steady states. We provide exact results pertaining to the entropy in the absence of many-body interactions but a working ansatz in their presence.

1.2 Background

One of our main questions is whether it is meaningful to attribute, to a nonequilibrium system, values of temperature or voltage which effectively describe systems in thermodynamic equilibrium. The question of a temperature measurement outside equilibrium has been asked by various researchers in a variety of contexts, classical and quantum, and has generated a vast amount of scientific discussion but with no meaningful consensus [1]. The first order of business is then to understand the notion of equilibrium and the role of temperature and voltage in the equilibrium setting. We do this briefly here and then highlight the complications in extending notions of equilibrium thermodynamics to the nonequilibrium setting.

Equilibrium is an axiomatic assumption in statistical mechanics and thermodynamics. Two systems are said to be in a state of mutual equilibrium if there is no net exchange of energy *and* matter between them. Two systems which are not in mutual equilibrium will evolve towards a state of mutual equilibrium in a manner dictated by the second law of thermodynamics, that is, by maximizing the sum of their entropies. An isolated system (of classical particles[1]) therefore evolves in a manner where parts of the system (subsystems) are in mutual equilibrium with each other thereby increasing the total entropy. An equilibrium system is one in which there is no net flow of energy or matter between parts of the system. A system in a state of complete internal equilibrium has a spatially uniform temperature.

The zeroth law of thermodynamics introduces the notion of temperature as a parameter labeling thermal equilibrium states. The existence of such equilibrium states is a central axiom of thermodynamics. Historically, the zeroth law of thermodynamics [2] appeared the latest but its importance lies in the pedagogical introduction of the notion of temperature. It states that thermal equilibrium states, labeled by their temperatures, form an equivalence class. That is, all systems labeled by the same temperature are in mutual *thermal* equilibrium. The zeroth law does not require the temperature to take on continuous values but only introduces temperature as a label (isotherm) for thermal equilibrium states. The equivalence property of thermal equilibrium states may be utilized to develop a thermometer: By choosing a system (thermometer) which is in mutual thermal equilibrium with the system of interest and having a readily measurable property which can be related directly to an empirical temperature. Properties which may be used to measure an empirical temperature are thermal expansion (e.g., of mercury), resistance of a metal, the thermopower of a thermocouple, etc.

---

[1]The entropy of an isolated quantum system is constant as shown explicitly in Chap. 5, Sect. 5.1.

   The intuitive concept of a "hotter" or "colder" temperature relies on the second law of thermodynamics. The Clausius statement of the second law of thermodynamics [3] establishes the direction of heat flow to be from the thermal equilibrium state at a hotter temperature to the one at colder temperature. A "hot" cup of coffee, for example, will cool down to reach the temperature of the room. The absolute temperature scale [4] which is used today was introduced by Thomson (Lord Kelvin) based on the Carnot's statement of the second law of thermodynamics. Thus, the notion of temperature draws from both the zeroth law and the second law of thermodynamics. Therefore, when defining a temperature for nonequilibrium systems, it becomes important to check its consistency with the laws of thermodynamics.

   The thermodynamics of equilibrium systems [5] deals with thermodynamic equilibrium states and the transformation between such states. A majority of the discussion restricts these transformations to be reversible. A reversible transformation is one where the change in the system is brought about so slowly that at each instant of time the system may be assumed to be in a state of internal equilibrium. A reversible process does not change the entropy of the universe and at any stage the entropy gained by the system may be returned to the environment and vice versa. A closed reversible cycle therefore does not change the entropy (or any other state function) at the end of the cycle.

   Early work on classical irreversible thermodynamics[2] considered situations where a local equilibrium could be assumed. In situations where such an assumption is valid, one may divide the system into smaller units with each such unit being close to equilibrium. The subunits are small enough to be considered homogenous and at equilibrium but large enough to warrant a macroscopic thermodynamic description. This leads to a major simplification and we may locally define temperature, chemical potential, and other thermodynamic variables in the usual way as it is done for equilibrium systems. As suggested by the title of this book, we consider problems that are far from equilibrium where the local equilibrium hypothesis is not valid. We restrict our attention to quantum fermion systems in a steady state.

1.3   Motivation

The need for the development of a thermodynamic theory of quantum systems far from equilibrium is more pressing than ever owing to the remarkable progress in the design and fabrication of nanoscale devices. Commercial electronic devices have undergone a steady miniaturization since the 1960s overtaking substantially Moore's initial prediction [8] that the trend would continue for a decade. The fabrication of mesoscopic devices in the 1980s has enhanced our understanding of electron transport at scales smaller than the mean free path of the electron

---

[2]The classic textbooks by de Groot [6] or Prigogine [7] provide a detailed exposition.

where Ohm's law is no longer valid [9]. The local equilibrium hypothesis would no longer be tenable at such length scales due to the absence of relaxation processes. This period saw the discovery of novel quantum phenomena such as conductance quantization [10], quantum Hall effect [11], and the experimental verification of the Aharonov–Bohm effect [12]. The period concurrently saw the development of scanning probe techniques [13–15] which revolutionized the measurement of local electronic properties, providing remarkable spatial resolutions. The measurement of temperature and voltage were soon to follow with the development of scanning probe thermometry [16] and scanning probe potentiometry [17], respectively. The theoretical basis for such measurements, however, has remained unclear and the present book clarifies some of these foundational issues.

Recent technological developments have enabled physicists to observe charge and heat transport in devices whose dimensions are several orders of magnitude smaller than the electron mean free path. Single-molecular junctions [18] and atomic contact junctions [19] represent the current frontier of miniaturization of electronic devices. Experiments are now able to measure the conductance and thermopower of single-molecular junctions [20] as well as atomic-size contact junctions [21]. While the quantization of electrical conductance in atomic contact junctions has been observed previously [22–24], recent years have seen a tremendous progress in the measurement of heat transport and dissipation at the atomic scale [25–27]. Figure 1.1 shows the experimental setup used by Cui et al. [26] to measure the heat current across a single-atom junction of gold and represents the current state-of-the-art in thermal measurements. Their setup is capable of measuring simultaneously the electrical and thermal conductances: For gold atomic contacts at room temperature, they find that the thermal conductance is quantized and follows the quantized electrical conductance. The insets (c) and (d) of Fig. 1.1 show the scanning electron micrographs of their experimental setup. The temperature of the probe tip is measured by monitoring the change in the resistance of the Pt resistance thermometer shown in Fig. 1.1 and the thermal resistance network shown in the inset (a) is used to calculate the thermal conductance.

The zeroth law of thermodynamics does not hold for systems outside equilibrium.[3] A thermometer records a temperature reading by reaching mutual equilibrium with the system of interest. In equilibrium, any two systems that have the same temperature reading will be in mutual equilibrium if brought into contact with each other. All degrees of freedom will have the same temperature for a system in a state of internal equilibrium. This would no longer be true outside equilibrium and the temperature that is measured would depend on the nature of interactions between the thermometer and the system. For example, a thermometer may couple strongly to some degrees of freedom while coupling weakly to others. There is no equipartition principle outside equilibrium and different degrees of freedom may have different temperatures. A thermometer may indeed measure the same

---

[3]Reference [1] provides a detailed exposition and reference list discussing the problems in extending the notion of temperature to nonequilibrium situations where no local equilibrium exists.

Experimental setup and scanning electron microscopy (SEM) images of the scanning thermal probes.

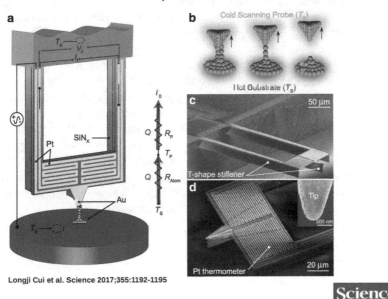

Longji Cui et al. Science 2017;355:1192-1195

Science

AAAS

Fig. 1.1 (**a**) Schematic of the experimental setup showing the Pt resistance thermometer which measures the temperature of the probe tip. The resistance of the thermometer is in turn measured by applying a sinusoidal current I_f and monitoring the voltage V_f. The electrical conductance is measured by the application of a small voltage across the single-atom junction and monitoring the resultant current. The thermal resistance network shown is used to calculate the thermal conductance of the junction by using the known value of the probe thermal resistance R_p. (**b**) depicts the formation of the atomic contact junction. (**c**) and (**d**) show the SEM images of the scanning probe. Reproduced with permission from [26]

temperature for two different systems that are out of equilibrium. However, the two systems, when brought into contact with each other, may not be in mutual equilibrium. The thermometer may couple strongly to some degrees of freedom in the first system while coupling strongly to others in the second system; there would be no reason to expect that the two systems would remain in mutual equilibrium when brought into contact with each other. We are well aware of this issue and we explicitly consider the electronic degrees of freedom in all the results presented in this thesis. However, the results are equally applicable to any fermionic degree of freedom so long as it is understood that one is talking about the temperature and chemical potential corresponding to those fermions alone.

The invalidity of the zeroth law outside equilibrium has prompted some authors to regard definitions of temperature or voltage as "operational" and lacking any fundamental meaning. For example, McLennan states in his book [28] "Clearly there is no zeroth law for a nonequilibrium situation. The reading of a thermometer may depend on its orientation, shape, surface properties and so on ... Nonequilibrium temperature is introduced for theoretical convenience rather than to take advantage of a basic principle." We disagree with this point of view. The question of the zeroth law of thermodynamics was analyzed by Stafford et al. in earlier works [29, 30] in some detail. The present book concerns itself mainly with the second law of thermodynamics. The results presented here take the notions of temperature and voltage well beyond their "operational" role in describing nonequilibrium systems.

One of the main results presented in this book is the proof of the Onsager's statement of the second law of thermodynamics. We shall see that this result provides a rigorous mathematical foundation for the notions of temperature and voltage in the nonequilibrium setting. Onsager's seminal work of 1931 [31, 32] introduced the so-called linear response coefficients and derived the reciprocal relations that they satisfy. Onsager's contribution was an influential starting point for the later development of nonequilibrium thermodynamics.

1.4 Formalism

The results presented in this book pertain to quantum fermion systems arbitrarily far from equilibrium. The theoretical basis for the description of quantum systems out of equilibrium traces back to the pioneering work of Martin, Schwinger [33, 34], Kadanoff, Baym [35], Keldysh [36], and others in the 1960s. This formalism is now referred to in literature as the nonequilibrium Green's function (NEGF) formalism and sometimes as the Keldysh formalism. The formalism was applied to the study of electron transport through an interacting nanostructure (e.g., Ref. [37]) during the 1990s. We do not review the NEGF formalism in the present book but include some useful results in the appendices. The reason for this is twofold. Firstly, a formal knowledge of Green's functions is not needed to understand the results presented in this book. Secondly, there are many excellent textbooks available on the subject which devote the proper time and space needed to develop the formalism.

We recommend the excellent book by Stefanucci and van Leeuwen [38]. In chapter 4 of Ref. [38], the authors develop the contour idea which is needed to deal with nonequilibrium problems. The contour idea can also be applied to the zero-temperature and finite-temperature (equilibrium) Green's function formalisms and has been conveniently presented in the same chapter. Chapter 4 ends with the derivation of the equation of motion on the contour for the general n-particle Green's function. These equations of motion, along with the boundary conditions [33, 39], completely specify the nonequilibrium many-body problem. If one could solve for the n-particle Green's function, one would be able to calculate any time-dependent ensemble average of any n-particle operator. The equations of motion, however, couple the n-particle Green's functions to the $(n-1)$ and $(n+1)$-particle Green's functions and one is tasked with solving this hierarchy of equations, which is discussed in detail in Chap. 5. In the absence of many-body interactions, the

n-particle Green's function only depends on the $(n-1)$-particle Green's function and the hierarchy can be solved exactly using Wick's theorem [40]. The interacting case can subsequently be treated perturbatively in the interparticle interaction strength.

A definite advantage of the authors' [38] presentation is that their contour formalism is presented in the most general terms and may be reduced to the appropriate Green's function formalism (e.g., the zero-temperature formalism, finite-temperature formalism, and the nonequilibrium Keldysh formalism) merely by an appropriate choice of the contour. Therefore, it serves as a pedagogical introduction to all the Green's function formalisms used in literature. Chapter 6 is devoted to the physical information contained in the single-particle Green's function. The lesser and greater Green's functions are introduced which contain information regarding the nonequilibrium particle and hole occupancies, respectively. We almost exclusively use the single-particle Green's functions in this book.[4] Chapters 4, 5, and 6 from Ref. [38] would provide more than sufficient background for the interested reader.

The starting point for our discussions is the Meir–Wingreen formula [37] for the particle current and, the analogous Bergfield–Stafford formula [41] for the heat current, through an interacting region for a quantum fermion system in a steady state. The Meir–Wingreen formula has been derived in the first section of chapter 16 in [38] and can be followed with the background from chapters 4, 5, and 6. However, in our view, it is not necessary for the reader to have this background. One may rather take the particle and heat current formulae in good faith and proceed further. The text is accompanied with physical interpretations for equations involving the Green's functions and therefore is accessible to a reader with a general knowledge of quantum mechanics.

1.5 Book Structure

The book is organized as follows. In Chap. 2, we present results which place temperature and voltage on a rigorous mathematical foundation. In Lemma 2.1 (2.2) we show that a voltage (temperature) measurement requires the specification of a temperature (voltage). These two results are shown to be connected to the Clausius statement of the second law of thermodynamics. In Theorems 2.1 and 2.2 we show that the uniqueness of the temperature and voltage measurement is a consequence of a different statement of the second law of thermodynamics. We refer to this form as the Onsager's statement of the second law (Theorem 2.1) due to its formal similarity with the positivity of the matrix of response coefficients as asserted (without proof)

---

[4]In Appendix B we encounter the irreducible kernel of the two-particle Green's function which is introduced in chapter 12 of Ref. [38]. Our discussion here pertains to the question of a noninvasive probe. We explicitly clarify, in Appendix B, the dependence of the currents on the probe-system interaction strength in the presence of arbitrary interactions within the system.

for consistency with the second law in Onsager's seminal work of 1931 [31, 32]. Our proof can be readily extended to prove the positivity of the matrix of linear response coefficients [31] for the case of quantum thermoelectric transport. Subsequently, we proceed to show in Theorem 2.3 that the solution always exists even in the case of a population inversion (Corollary 2.3.1) for which one obtains negative (absolute) temperature solutions. One may obtain negative temperatures if the quantum system has a spectrum that is bounded above, which we illustrate for a two-level system. These results are very general: They apply to any quantum fermion system in a steady state, arbitrarily far from equilibrium, with arbitrary many-body interactions within the quantum system. In the presence of many-body interactions, we show our results for noninvasive probes. In the absence of many-body interactions, we find that our results hold for arbitrarily strong probe-system couplings as shown in Appendix B. A detailed discussion of what constitutes an ideal probe is also central to our analysis and has been covered in Chap. 2 as well as in Appendix B.

Chapter 3 considers a question motivated by the third law of thermodynamics: What is the coldest possible temperature for a nonequilibrium quantum system? We note at the very beginning of the chapter that absolute zero cannot be obtained as a measurement outcome and then proceed to discuss a nonequilibrium setting where one reservoir is held close to absolute zero and the other at finite temperature, a problem outside the scope of linear response theory. We consider explicitly the case where the transmission function between the probe and the finite-temperature reservoir has a node and derive analytical expressions for the probe temperature. The probe temperature for this case obeys a polynomial equation of the order dictated by the transmission node and we explicitly see that such polynomials must have a unique real root, as expected from the results of Chap. 2. We present numerical results for several single-molecular junctions and find excellent agreement with the analytical results. The third law of thermodynamics is also discussed extensively in Chap. 5 on entropy.

In Chap. 4, we present a novel experimental method for temperature measurements using a scanning tunneling microscope whose working principle is closely related to the findings in Chap. 2. Since we found in Chap. 2 that temperature and voltage must be measured simultaneously to ensure uniqueness, we highlight the need to amend earlier notions [42] of thermometry and potentiometry. We show that in nanoscale conductors where the Wiedemann–Franz law [43] is valid, temperature measurements can be made in the tunneling regime. This proposed *scanning tunneling thermometer* would be capable of mapping sub-angstrom temperature variations and, if implemented practically, would constitute a remarkable improvement over the existing state-of-the-art in scanning probe thermometry at the nanoscale whose spatial resolution is currently around $\sim$7 nm.

Any theory of thermodynamics is incomplete without a thorough exploration of the notion of entropy. In Chap. 5, we discuss the entropy in the context of the steady-state quantum transport problem. We provide an expression for the exact entropy of a quantum system which, however, relies on the knowledge of the scattering states which would not be fully available to a local observer. We formulate the entropy inferred by a local observer and develop a hierarchy of inequalities for the entropies.

The entropy inequalities that we obtain have an intimate relationship with the information available to an observer from a local measurement: More information implies a smaller entropy. The analysis in this chapter is done in the scattering basis and we present rigorous results for the noninteracting case. Including interactions will cause the scattering states to mix and the analysis becomes considerably more complicated. We then straightforwardly extend the definition to apply for such a case but, since a rigorous justification is missing, we would like to call it an ansatz when interactions are present. We also discuss the use of entropy as a metric to quantify the departure from equilibrium, when it is appropriately normalized. We also present proofs of the third law of thermodynamics for generic open quantum systems. We show that fully quantum open systems (having no localized states) will have vanishing entropy at zero temperature whereas generic open quantum systems with localized states may, extremely rarely,[5] have finite contributions due to such localized states.

Some useful results pertaining to the Green's functions are included in the appendices. In order to benefit the reader, we have included brief introductions at the start of every chapter and the reader may choose to skip to a chapter of particular interest, hopefully, without loss of context.[6]

References

1. J. Casas-Vázquez, D. Jou, Rep. Prog. Phys. **66**(11), 1937 (2003). http://stacks.iop.org/0034-4885/66/i=11/a=R03
2. R. Fowler, E. Guggenheim, *Statistical Thermodynamics* (Cambridge University Press, Cambridge, 1939)
3. R. Clausius, Ann. Phys. **169**(12), 481 (1854). https://doi.org/10.1002/andp.18541691202. http://dx.doi.org/10.1002/andp.18541691202
4. W. Thomson, *Mathematical and Physical Papers* (Cambridge University Press, Cambridge, 1848), pp. 100–106
5. F. Reif, *Fundamentals of Statistical and Thermal Physics* (McGraw-Hill, New York, 1965)
6. S. De Groot, P. Mazur, *Non-equilibrium Thermodynamics* (North-Holland, Amsterdam, 1962)
7. P. Glansdorff, I. Prigogine, *Thermodynamic Theory of Structure, Stability and Fluctuations* (Wiley, New York, 1971)
8. G.E. Moore, Electronics **38**(8), 114–117 (1965)
9. S. Datta, *Electronic Transport in Mesoscopic Systems* (Cambridge University Press, Cambridge, 1995)
10. B.J. van Wees, H. van Houten, C.W.J. Beenakker, J.G. Williamson, L.P. Kouwenhoven, D. van der Marel, C.T. Foxon, Phys. Rev. Lett. **60**, 848 (1988). https://doi.org/10.1103/PhysRevLett.60.848. https://link.aps.org/doi/10.1103/PhysRevLett.60.848
11. K.v. Klitzing, G. Dorda, M. Pepper, Phys. Rev. Lett. **45**, 494 (1980). https://doi.org/10.1103/PhysRevLett.45.494. https://link.aps.org/doi/10.1103/PhysRevLett.45.494

---

[5]Mathematically, we show that the entropy vanishes almost everywhere on the real line except for a finite number of points (having zero measure) which correspond to the localized states.

[6]The analysis in Chap. 3 follows from the more general results presented in Chap. 2 and we provide a briefer introduction.

12. S. Washburn, R.A. Webb, Adv. Phys. **35**, 375 (1986)
13. G. Binnig, H. Rohrer, C. Gerber, E. Weibel, Appl. Phys. Lett. **40**(2), 178 (1982). http://dx.doi.
    org/10.1063/1.92999. http://scitation.aip.org/content/aip/journal/apl/40/2/10.1063/1.92999
14. G. Binnig, H. Rohrer, Surf. Sci. **126**(1), 236 (1983). http://dx.doi.org/10.1016/0039-
    6028(83)90716-1. http://www.sciencedirect.com/science/article/pii/0039602883907161
15. G. Binnig, C.F. Quate, C. Gerber, Phys. Rev. Lett. **56**, 930 (1986). https://doi.org/10.1103/
    PhysRevLett.56.930. http://link.aps.org/doi/10.1103/PhysRevLett.56.930
16. C.C. Williams, H.K. Wickramasinghe, Appl. Phys. Lett. **49**(23), 1587 (1986). http://dx.doi.org/
    10.1063/1.97288. http://scitation.aip.org/content/aip/journal/apl/49/23/10.1063/1.97288
17. P. Muralt, D.W. Pohl, Appl. Phys. Lett. **48**(8), 514 (1986). http://dx.doi.org/10.1063/1.96491.
    http://scitation.aip.org/content/aip/journal/apl/48/8/10.1063/1.96491
18. J.C. Cuevas, E. Scheer, *Molecular Electronics: An Introduction to Theory and Experiment*
    (World Scientific, Singapore, 2010)
19. J.K. Gimzewski, R. Möller, Phys. Rev. B **36**, 1284 (1987). https://doi.org/10.1103/PhysRevB.
    36.1284. https://link.aps.org/doi/10.1103/PhysRevB.36.1284
20. J.R. Widawsky, P. Darancet, J.B. Neaton, L. Venkataraman, Nano Lett. **12**(1), 354 (2012).
    https://doi.org/10.1021/nl203634m. http://dx.doi.org/10.1021/nl203634m. PMID: 22128800
21. C. Evangeli, M. Matt, L. Rincn-Garca, F. Pauly, P. Nielaba, G. Rubio-Bollinger, J.C. Cuevas,
    N. Agrat, Nano Lett. **15**(2), 1006 (2015). https://doi.org/10.1021/nl503853v. http://dx.doi.org/
    10.1021/nl503853v. PMID: 25607343
22. N. Agraït, J.G. Rodrigo, S. Vieira, Phys. Rev. B **47**, 12345 (1993)
23. E. Scheer, N. Agraït, J.C. Cuevas, A.L. Yeyati, B. Ludoph, A. Martín-Rodero, G.R. Bollinger,
    J.M. van Ruitenbeek, C. Urbina, Nature **394**, 154 EP (1998). http://dx.doi.org/10.1038/28112
24. N. Agraït, A. Levy Yeyati, J.M. van Ruitenbeek, Phys. Rep. **377**, 81 (2003, and references
    therein)
25. W. Lee, K. Kim, W. Jeong, L.A. Zotti, F. Pauly, J.C. Cuevas, P. Reddy, Nature **498**(7453), 209
    (2013). http://dx.doi.org/10.1038/nature12183
26. L. Cui, W. Jeong, S. Hur, M. Matt, J.C. Klöckner, F. Pauly, P. Nielaba, J.C. Cuevas,
    E. Meyhofer, P. Reddy, Science **355**(6330), 1192 (2017). https://doi.org/10.1126/science.
    aam6622. http://science.sciencemag.org/content/355/6330/1192
27. N. Mosso, U. Drechsler, F. Menges, P. Nirmalraj, S. Karg, H. Riel, B. Gotsmann, Nat.
    Nanotechnol. **12**(5), 430 (2017). http://dx.doi.org/10.1038/nnano.2016.302. Letter
28. J. McLennan, *Introduction to Non-equilibrium Statistical Mechanics* (Prentice Hall, Upper
    Saddle River, 1989)
29. J. Meair, J.P. Bergfield, C.A. Stafford, P. Jacquod, Phys. Rev. B **90**, 035407 (2014). https://doi.
    org/10.1103/PhysRevB.90.035407. http://link.aps.org/doi/10.1103/PhysRevB.90.035407
30. C.A. Stafford, Phys. Rev. B **93**, 245403 (2016). https://doi.org/10.1103/PhysRevB.93.245403.
    http://link.aps.org/doi/10.1103/PhysRevB.93.245403
31. L. Onsager, Phys. Rev. **37**, 405 (1931). https://doi.org/10.1103/PhysRev.37.405. http://link.aps.
    org/doi/10.1103/PhysRev.37.405
32. L. Onsager, Phys. Rev. **38**, 2265 (1931). https://doi.org/10.1103/PhysRev.38.2265. https://link.
    aps.org/doi/10.1103/PhysRev.38.2265
33. P.C. Martin, J. Schwinger, Phys. Rev. **115**, 1342 (1959). https://doi.org/10.1103/PhysRev.115.
    1342. https://link.aps.org/doi/10.1103/PhysRev.115.1342
34. J. Schwinger, J. Math. Phys. **2**(3), 407 (1961). https://doi.org/10.1063/1.1703727. https://doi.
    org/10.1063/1.1703727
35. L.P. Kadanoff, G. Baym, *Quantum Statistical Mechanics* (Addison Wesley, Boston, 1994)
36. L.V. Keldysh, Zh. Eksp. Teor. Fiz. **47**, 1515 (1964)
37. Y. Meir, N.S. Wingreen, Phys. Rev. Lett. **68**, 2512 (1992)
38. G. Stefanucci, R. van Leeuwen, *Nonequilibrium Many-Body Theory Of Quantum Systems: A
    Modern Introduction* (Cambridge University Press, Cambridge, 2013)
39. R. Kubo, J. Phys. Soc. Jpn. **12**(6), 570 (1957). https://doi.org/10.1143/JPSJ.12.570. https://doi.
    org/10.1143/JPSJ.12.570

40. G.C. Wick, Phys. Rev. **80**, 268 (1950). https://doi.org/10.1103/PhysRev.80.268. https://link.aps.org/doi/10.1103/PhysRev.80.268
41. J.P. Bergfield, C.A. Stafford, Nano Lett. **9**, 3072 (2009)
42. H.L. Engquist, P.W. Anderson, Phys. Rev. B **24**, 1151 (1981). https://doi.org/10.1103/PhysRevB.24.1151. http://link.aps.org/doi/10.1103/PhysRevB.24.1151
43. N.W. Ashcroft, N.D. Mermin, *Solid State Physics* (Brooks/Cole - Thomson Learning, Pacific Grove, 1976)

Chapter 2
Temperature and Voltage

"The law that entropy always increases holds, I think, the supreme position among the laws of Nature. If someone points out to you that your pet theory of the universe is in disagreement with Maxwell's equations—then so much the worse for Maxwell's equations. If it is found to be contradicted by observation—well, these experimentalists do bungle things sometimes. But if your theory is found to be against the second law of thermodynamics I can give you no hope; there is nothing for it but to collapse in deepest humiliation."—Arthur Eddington [1].

2.1 Introduction

Temperature and voltage[1] are basic thermodynamic quantities which are routinely measured in all sorts of experiments. The definition of temperature and voltage are well established in equilibrium. The zeroth law encodes the definition of temperature as an equivalence class of equilibrium systems where any two systems from the same class, when brought into contact, do not have a net exchange of energy. The second law tells us that the energy (in the form of heat) flows from a system at a higher temperature to a system at lower temperature in the absence of work. Similarly, particles flow from a system at higher chemical potential to a system at lower chemical potential in the absence of temperature gradients. Here we show that such formulations of the second law are indeed possible for quantum systems far from equilibrium and that they have profound consequences for the measurement of temperature and chemical potential (or voltage when dealing with

---

[1] Voltage here refers to the electrochemical potential (μ) and not just the electrostatic potential (V). A voltmeter in fact measures the electrochemical potential difference which of course includes the electrostatic contribution $\mu = \mu_0 + eV$.

© Springer Nature Switzerland AG 2019
A. Shastry, *Theory of Thermodynamic Measurements of Quantum Systems Far from Equilibrium*, Springer Theses, https://doi.org/10.1007/978-3-030-33574-8_2

charged particles). Just as Eddington notes, our statements of the second law are not dependent on the nature of interactions within the quantum system.

The definition of thermodynamic variables in a system far from equilibrium is of fundamental interest. It serves as a necessary step towards the construction of nonequilibrium thermodynamics which has gained renewed interest in recent years [2–9]. On the experimental side, scanning tunneling potentiometry [10] (i.e., the measurement of local voltages) has undergone tremendous progress over the past two decades [11–13] achieving sub-angstrom spatial resolutions. Scanning thermometry [14] has proven significantly more challenging but is currently undergoing a rapid evolution toward nanometer resolution [15–18]. The progress on the theoretical side is therefore ever more important since predictions will soon be experimentally testable. Unfortunately, even the notions of temperature and voltage out of equilibrium have not been thought out carefully. Until now, mainly operational definitions [6, 19–28] have been advanced leading to a competing panoply of often contradictory definitions of such basic observables as temperature and voltage. The results in this chapter fill this gap by developing a rigorous mathematical description of local thermodynamic measurements.

One other point deserves mention before we look at the results of this chapter. Out of equilibrium, different degrees of freedoms can have different temperatures. We know this fact from NMR spectroscopy where the nuclear spin temperature can be negative but the atom or electron temperature is clearly positive. Another example is that of a laser, where electron temperature is negative due to population inversion but the lattice/gas temperature is positive. We therefore have to clearly demarcate between lattice temperature [29–31], photon temperature [32–34], and electron temperature [6, 7, 19, 22, 24, 26, 27, 35]. Our work concerns itself only with the electron temperature. The results presented in this chapter are, however, true for any fermionic system, charged or neutral.

The main results of this chapter are presented in the lemmas and theorems in the subsequent sections. The brief summary of these results is as follows. We show that temperature and voltage cannot be measured separately for a system out of equilibrium. This is shown in Lemmas 2.1, 2.2 and Theorems 2.1 and 2.2. Lemma 2.1 shows that a voltage-only measurement is nonunique and hence voltage is ill-defined without an accompanying temperature measurement. Lemma 2.2 shows that a temperature-only measurement is nonunique and hence temperature is ill-defined without an accompanying voltage measurement. Theorem 2.1 is in fact the Onsager statement of the second law which, to our knowledge, is proved for the first time. Lemmas 2.1 and 2.2 are also statements of the second law related to the Clausius statement. Theorem 2.2 shows that the simultaneous measurement is unique and hence the only meaningful way to define temperature and voltage. Theorem 2.3 answers the question of the existence of a solution. Theorem 2.3 and its Corollary 2.3.1 establish that a solution always exists. There is also the question of an ideal probe which is addressed extensively.

The results here are quite general in their applicability: They hold for steady-state quantum systems arbitrarily far from equilibrium, with arbitrary interactions within the quantum system, and for any fermionic system, charged or neutral.

One important consideration is that of a noninvasive probe. In the presence of interactions, our results only hold for the noninvasive probe limit. Analysis outside this limit appears to be mathematically complicated. If interactions are absent, then the noninvasive probe limit is not needed and the results hold for arbitrarily strong system-probe couplings. This latter point is explicitly shown in Appendix B, Sect. B.2.

2.2 Expression for the Currents

We use the nonequilibrium Green's function formalism (NEGF) for describing the motion of electrons within a quantum conductor. The starting point for our discussions is the Meir–Wingreen formula [36] for the particle current and, the analogous Bergfield–Stafford formula [37] for the heat current, through an inter-acting region for a quantum fermion system in a steady state. We write the two formulae in a combined form below [7] using the index $v = \{0, 1\}$ to refer to the two cases, respectively. The most general expression for the nonequilibrium steady-state particle current ($v = 0$) [36] and the electronic contribution to the heat current ($v = 1$) [37] flowing into a macroscopic electron reservoir P is

$$I_p^{(v)} = -\frac{i}{h} \int_{-\infty}^{\infty} d\omega (\omega - \mu_p)^v \, \mathrm{Tr} \left\{ \Gamma^P(\omega) \left(G^<(\omega) + f_p(\omega) \left[G^>(\omega) - G^<(\omega) \right] \right) \right\},$$

(2.1)

where $\Gamma^P(\omega)$ is the tunneling-width matrix describing the coupling of the probe to the system and

$$f_p(\omega) = \left\{ 1 + \exp\left[(\omega - \mu_p)/k_B T_p \right] \right\}^{-1}$$

is the Fermi–Dirac distribution of the probe. $G^<(\omega)$ and $G^>(\omega)$ are the Fourier transforms of the Keldysh "lesser" and "greater" Green's functions [38], describing the nonequilibrium electron and hole distributions within the system, respectively.

Equation (2.1) is a general result valid for arbitrary interactions within the quantum system. It is also valid for situations arbitrarily far from equilibrium. A lot of insight can be gained by a rearrangement of Eq. (2.1) into the Landauer–Büttiker form [7]

$$I_p^{(v)} = \frac{1}{h} \int_{-\infty}^{\infty} d\omega (\omega - \mu_p)^v \mathcal{T}_{ps}(\omega) [f_s(\omega) - f_p(\omega)].$$

(2.2)

The Landauer–Büttiker form was originally derived for noninteracting electrons within the scattering approach; therefore, it is crucial to note that Eq. (2.2) is valid for arbitrary interactions. It is simply a rearrangement of Eq. (2.1).

Equation (2.2) is describing the rate of particle and heat exchanged by the macroscopic probe reservoir p (which is taken to be at equilibrium) with the nonequilibrium system of interest s. We show that $\mathcal{T}_{ps}(\omega)$ satisfies the property of a transmission function and is non-negative. We show also that $f_s(\omega)$ satisfies the property of a fermionic distribution function and takes on values in the range $[0, 1]$.

We first write down the expression for f_s and $\mathcal{T}_{ps}$ in terms of the Green's functions. We refer to f_s as the local nonequilibrium distribution function of the system, as sampled by the probe [7]

$$f_s(\omega) \equiv \frac{\text{Tr}\{\Gamma^p(\omega)G^<(\omega)\}}{2\pi i\,\text{Tr}\{\Gamma^p(\omega)A(\omega)\}}, \qquad (2.3)$$

and the effective probe-system transmission function

$$\mathcal{T}_{ps}(\omega) = 2\pi\,\text{Tr}\{\Gamma^p(\omega)A(\omega)\}. \qquad (2.4)$$

$A(\omega)$ is the spectral function given by

$$A(\omega) = \frac{1}{2\pi i}\left(G^<(\omega) - G^>(\omega)\right), \qquad (2.5)$$

and is Hermitian and positive-semidefinite as shown in Appendix A.

Since the probe-system coupling $\Gamma^p(\omega)$ is also positive-semidefinite (due to causality [38]), we note that

$$\begin{aligned}
\text{Tr}\{A(\omega)\Gamma(\omega)\} &= \text{Tr}\left\{A(\omega)^{1/2}A^{1/2}(\omega)\Gamma(\omega)\right\} \\
&= \text{Tr}\left\{A^{1/2}(\omega)\Gamma(\omega)A^{1/2}(\omega)\right\} \\
&\geq 0,
\end{aligned} \qquad (2.6)$$

where $A^{1/2}(\omega)$ is the positive-semidefinite square root of $A(\omega)$. $A^{1/2}(\omega)\Gamma(\omega)A^{1/2}(\omega)$ becomes positive-semidefinite when $A^{1/2}(\omega)$ and $\Gamma(\omega)$ are positive-semidefinite [39] and therefore we have

$$\mathcal{T}_{ps}(\omega) \geq 0, \ \forall \omega \in \mathbb{R}. \qquad (2.7)$$

We also show in Appendix A that f_s satisfies the property of a fermionic distribution function:

$$0 \leq f_s(\omega) \leq 1 \ \forall \omega \in \mathbb{R}. \qquad (2.8)$$

We start our analysis with the following postulate and explain its physical significance.

Postulate 2.1 The local probe-system transmission function $\mathcal{T}_{ps} : \mathbb{R} \rightarrow [0, \infty)$ and the nonequilibrium distribution function $f_s : \mathbb{R} \rightarrow [0, 1]$ are measurable over any interval $[a, b] \in \mathbb{R}$ and $\mathcal{T}_{ps}(\omega)$ satisfies

$$0 < \int_{-\infty}^{\infty} d\omega \mathcal{T}_{ps}(\omega) < \infty \tag{2.9}$$

and

$$\left| \int_{-\infty}^{\infty} d\omega \, \omega \mathcal{T}_{ps}(\omega) \right| < \infty. \tag{2.10}$$

The measurability of $\mathcal{T}_{ps}(\omega)$ and $f_s(\omega)$ is taken to lend meaning to the currents in Eq. (2.2). We point out that the finiteness of the two integrals given in Eqs. (2.9) and (2.10) is more relevant to our discussion of existence of solutions in Sect. 2.5. Our result on uniqueness, as stated in Theorem 2.2, is somewhat stronger and requires only that the function $\mathcal{T}_{ps}(\omega)$ grows slower than exponentially for large values of energy (for $\omega \to \pm\infty$).

On physical grounds, the probe-sample transmission function $\mathcal{T}_{ps}(\omega)$ can be argued to have a compact support (non-zero only for some finite interval $[\omega_-, \omega_+] \subset \mathbb{R}$). It is easy to see that $\mathcal{T}_{ps}$ must have a lower bound ω_- such that $\mathcal{T}_{ps}(\omega) = 0 \; \forall \, \omega < \omega_-$, since physical Hamiltonians must have a finite ground-state energy. However, for energies larger than the probe work function (ω_+), it can be argued that the particle will merely pass through the probe and not contribute to the steady-state currents into the probe. $\mathcal{T}_{ps}(\omega)$ then has a compact support and satisfies Eqs. (2.9) and (2.10). In Sect. 2.5, we comment upon the limiting case where the measure of $\omega \mathcal{T}_{ps}(\omega)$ in Eq. (2.10) tends to infinity. The absolute value on the l.h.s. in Eq. (2.10) is somewhat redundant since the limiting case must have $\omega_+ \to \infty$ while $\omega_- \to -\infty$ is ruled out based on the principle that any physical spectrum has a finite ground-state energy. We note that Eqs. (2.9) and (2.10) also imply

$$0 < \int_{-\infty}^{\infty} d\omega \mathcal{T}_{ps}(\omega) f_s(\omega), \int_{-\infty}^{\infty} d\omega \mathcal{T}_{ps}(\omega) f_p(\omega) < \infty \tag{2.11}$$

and

$$\int_{-\infty}^{\infty} d\omega \, \omega \mathcal{T}_{ps}(\omega) f_s(\omega), \int_{-\infty}^{\infty} d\omega \, \omega \mathcal{T}_{ps}(\omega) f_p(\omega) < \infty. \tag{2.12}$$

2.3 Local Measurements

The local voltage and temperature of a nonequilibrium quantum system, as measured by a scanning thermoelectric probe, is defined by the simultaneous conditions of vanishing net charge dissipation *and* vanishing net heat dissipation into the probe [7, 9, 26, 27, 35, 40]:

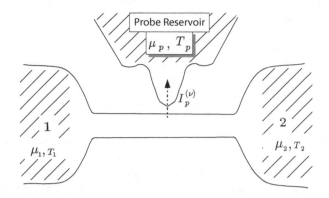

Fig. 2.1 Illustration of the measurement setup: The quantum conductor represented above is in a nonequilibrium steady state. A weakly coupled scanning tunneling probe noninvasively measures the local voltage (μ_p) and local temperature (T_p) *simultaneously*: By requiring both a vanishing net charge exchange ($I_p^{(0)} = 0$) *and* a vanishing net heat exchange ($I_p^{(1)} = 0$) with the system. The nonequilibrium steady state has been prepared, in this particular illustration, via the electrical and thermal bias of the strongly coupled reservoirs (1 and 2). The measurement method itself is completely general and does not depend upon (a) how such a nonequilibrium steady state is prepared, (b) how far from equilibrium the quantum electron system is driven, and (c) the nature of interactions within that system

$$I_p^{(\nu)} = 0, \quad \nu \in \{0, 1\}, \tag{2.13}$$

where $\nu = 0, 1$ correspond to the electron number current and the electronic contribution to the heat current, respectively. Equation (2.13) gives the conditions under which the probe is in local equilibrium with the sample, which is itself arbitrarily far from equilibrium.

We define the system's local temperature and voltage using a probe that is weakly coupled via a tunnel barrier. The other end of this scanning probe [41, 42] is the macroscopic electron reservoir whose temperature and voltage are both adjusted (Fig. 2.1) until Eq. (2.13) is satisfied. A weakly coupled probe is a useful theoretical construction for our analysis and the extension of our results beyond the weak-coupling limit is an open question. We explain the physical basis of weak coupling in Sect. 2.3.2 and derive some useful formulae. Before proceeding to the main results, we take a short detour which will help us better understand the physical meaning behind the quantities we defined here and the measurement condition of Eq. (2.13). We illustrate their meaning by considering an example system below.

2.3.1   Example: Anthracene Junction

We introduced the nonequilibrium spectrum f_s and the probe-system transmission function $\mathcal{T}_{ps}$ in Eqs. (2.3) and (2.4), respectively. It is beneficial to have an intuitive picture of these quantities. When the transport is dominated by elastic processes, the local probe-system transmission function $\mathcal{T}_{ps}$ can be thought of as the sum of the probe-reservoir transmission functions over all the reservoirs. Similarly, the local nonequilibrium distribution function becomes a linear combination of the Fermi functions corresponding to the different reservoirs (see also Appendix C).

The local measurement condition of Eq. (2.13) lends itself to the easiest physical interpretation when the probe is in the broadband limit. We discuss this limit in detail in Sect. 2.5.2 on ideal probes. Succinctly put, the measurement condition just becomes the following statement. We look for an equilibrium system with the distribution f_p which gives us the same average energy and average occupancy as that of the nonequilibrium system with the distribution f_s. The spectrum and distribution function of the system are sampled locally by the probe and are experimentally accessible [43].

We consider an anthracene molecular junction driven out of equilibrium by both temperature and voltage biases as shown in Fig. 2.2. The bias conditions were taken to be $T_1 = 100\,\mathrm{K}$ (reservoir 1 is indicated with blue squares), $T_2 = 300\,\mathrm{K}$ (reservoir 2 is indicated with red squares), and $\mu_2 - \mu_1 = 0.2\,\mathrm{eV}$. In the upper panel of Fig. 2.2, we show the voltage and temperature distributions along the molecule as measured by a scanning probe whose tip is 3.5 Å above the plane of the molecule.

There are two representative points on the molecule which are shown in the lower panel of Fig. 2.2. Point 2 illustrates the measurement condition very

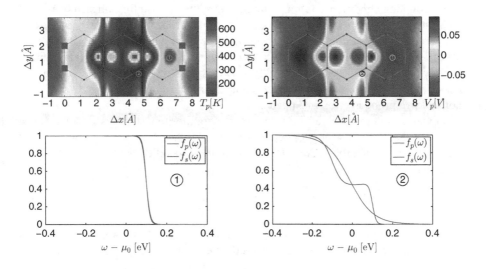

Fig. 2.2 Upper panel shows the temperature (left) and voltage (right) variations along the molecular junction as seen by a probe scanning at a height of 3.5 Å. Lower panel shows the nonequilibrium distribution of the system f_s (red) and the equilibrium Fermi–Dirac distribution f_p (blue) of the probe for two representative points marked 1 and 2. The measurement finds f_p so that the mean energy and occupancy of the probe is the same as that of the nonequilibrium system locally. The distribution f_s at point 2 has the clear feature of mixing of two Fermi–Dirac distributions corresponding to the two reservoirs (shown with red and blue squares in the upper panel)

clearly. One can see that the distribution function for point 2 has two clear steps corresponding to different the chemical potentials of the two reservoirs. The two steps also have different widths corresponding to the different temperatures of the two reservoirs. The measurement idea is then the following: We want to fit a Fermi–Dirac distribution such that it gives the same local occupancy and energy of the nonequilibrium system with the distribution f_s (see also the discussion in Ref. [7]). In other words, we want to find an equilibrium distribution f_p which gives the same zeroth and first moments of the system's local energy distribution[2] This corresponds to the conditions of zero particle and heat currents as written by the measurement condition in Eq. (2.13). The broadband condition has important implications for measurements which we discuss in Sect. 2.5 on the existence of solutions in the present chapter. We refer the reader also to the discussion in Ref. [7].

2.3.2 Noninvasive Measurements

When the coupling of the probe to the system is weak, we may take $\mathcal{T}_{ps}(\omega)$ in Eq. (2.4) and the local nonequilibrium distribution function $f_s(\omega)$ to be independent of the probe temperature T_p and chemical potential μ_p. While both $\mathcal{T}_{ps}(\omega)$ and $f_s(\omega)$ depend upon the local probe-system coupling in an obvious manner, the weak-coupling condition essentially implies that the nonequilibrium steady state of the system is unperturbed by the introduction of the probe terminal. The voltage and temperature of the probe itself play no role in preparing the nonequilibrium steady state. In other words, the probe does *not* drive the system but merely exchanges energy and particles across a weakly coupled tunnel barrier and constitutes a *noninvasive* measurement. A precise analysis of the conditions necessary for a noninvasive probe is given in Appendix B.

Given a system prepared in a certain nonequilibrium steady state (e.g., by a particular bias of the strongly coupled reservoirs), the currents given by Eq. (2.2) are functions of the probe Fermi–Dirac distribution specified by its temperature and chemical potential

$$I_p^{(\nu)} \equiv I_p^{(\nu)}(\mu_p, T_p). \tag{2.14}$$

It can be seen that the currents are continuous functions of $\mu_p \in (-\infty, \infty)$ and $T_p \in (0, \infty)$ with continuous gradient vector fields defined by

$$\nabla I_p^{(\nu)} \equiv \left(\frac{\partial I_p^{(\nu)}}{\partial \mu_p}, \frac{\partial I_p^{(\nu)}}{\partial T_p} \right). \tag{2.15}$$

---

[2]This is further made clear in Sect. 2.5.2 [see Eq. (2.56)]. The local spectrum introduced in Sect. 2.5.2 is proportional to the probe-system transmission function in the broadband limit.

With k_B set to unity, we compute the gradients of the currents using Eq. (2.2). We find the gradient of the number current to be

$$\nabla I_p^{(0)} = \left(- \mathcal{L}_{ps}^{(0)}, -\frac{\mathcal{L}_{ps}^{(1)}}{T_p} \right). \tag{2.16}$$

The gradient of the heat current reduces to

$$\nabla I_p^{(1)} = \left(- \mathcal{L}_{ps}^{(1)} - I_p^{(0)}, -\frac{\mathcal{L}_{ps}^{(2)}}{T_p} \right), \tag{2.17}$$

where we define the response coefficients $\mathcal{L}_{ps}^{(\nu)}$ as

$$\begin{aligned}
\mathcal{L}_{ps}^{(\nu)} &\equiv \mathcal{L}_{ps}^{(\nu)}(\mu_p, T_p) \\
&= \frac{1}{h} \int_{-\infty}^{\infty} d\omega (\omega - \mu_p)^{\nu} \mathcal{T}_{ps}(\omega) \left(-\frac{\partial f_p}{\partial \omega} \right),
\end{aligned} \tag{2.18}$$

which are easily seen to be finite.[3] Expressions (2.16) and (2.17) for the current gradients are valid for a *noninvasive* probe measurement for steady-state transport arbitrarily far from equilibrium and with arbitrary interactions within the quantum system.

Although the coefficients $\mathcal{L}_{ps}^{(\nu)}$ formally resemble the Onsager linear response coefficients [44] of an elastic quantum conductor [45], it is very important to note that we do *not* make the assumptions of linear response, time-reversal symmetry, local equilibrium, or elastic transport in the above definition of $\mathcal{L}_{ps}^{(\nu)}$: The system itself may be arbitrarily far from equilibrium with arbitrary inelastic scattering processes. The coefficients above appear naturally when we calculate the gradient fields defined by Eq. (2.15) and the gradient operator is of course given by the first derivatives. Our main results follow from an analysis of the properties of these gradient fields.

2.4   Uniqueness and the Second Law

We now turn to one of the central problems which we set out to address: $I_p^{(\nu)}(\mu_p, T_p) = 0$, with $\nu = \{0, 1\}$, is a system of coupled nonlinear equations in two variables that defines our local voltage and temperature measurement. There is no *a priori* reason to expect a unique solution even if a solution exists. We begin the section with statements of the second law of thermodynamics and conclude by showing that the uniqueness of the measurement emerges as a consequence.

---

[3] $\mathcal{L}_{ps}^{(\nu)}(\mu_p, T_p)$ are finite even if $\mathcal{T}_{ps}(\omega)$ and $\omega \mathcal{T}_{ps}(\omega)$ do not obey the finite measure conditions of Postulate 2.1 due to the exponentially decaying tails of the Fermi-derivative. We merely need $\mathcal{T}_{ps}(\omega)$ to grow slower than exponentially for $\omega \to \pm\infty$.

2.4.1 Statements of the Second Law

We note that $\forall \, \mu_p \in (-\infty, \infty)$ and $T_p \in (0, \infty)$,

$$\mathcal{L}_{ps}^{(0)}(\mu_p, T_p) > 0$$

$$\mathcal{L}_{ps}^{(2)}(\mu_p, T_p) > 0, \tag{2.19}$$

since $\mathcal{T}_{ps}(\omega) \geq 0$, and the measure of $\mathcal{T}_{ps}(\omega)$ and the Fermi-function derivative are both non-zero and strictly positive. This leads to two statements of the second law of thermodynamics, related to the Clausius statement, which are presented in the following two lemmas. The idea is to choose the correct contour for each case and evaluate the line integral over the current gradients given by Eqs. (2.16) and (2.17). A cursory glance at the number current gradient in Eq. (2.16) suggests that the contour should be defined over a constant temperature while the heat current gradient in Eq. (2.17) suggests a line integral over a constant voltage contour.

Lemma 2.1 *The number current contour defined by* $I_p^{(0)}(\mu_p, T_p) = 0$ *exists for all* $T_p \in (0, \infty)$ *and defines a function* $M : (0, \infty) \to \mathbb{R}$, *where* $\mu_p = M(T_p)$, *such that the second law of thermodynamics is obeyed:*

$$I_p^{(0)}(\mu_p', T_p) > 0, \text{ if } \mu_p' < \mu_p \text{ and}$$

$$I_p^{(0)}(\mu_p', T_p) < 0, \text{ if } \mu_p' > \mu_p. \tag{2.20}$$

Proof We first show that $I^{(0)}(\mu_p, T_p) = 0$ is satisfied for all $T_p \in (0, \infty)$. For any $T_p \in (0, \infty)$, we have

$$\lim_{\mu_p \to -\infty} I^{(0)}(\mu_p, T_p) = \frac{1}{h} \int_{-\infty}^{\infty} d\omega \mathcal{T}_{ps}(\omega) \left[f_s(\omega) - \lim_{\mu_p \to -\infty} f_p(\omega) \right]$$

$$= \frac{1}{h} \int_{-\infty}^{\infty} d\omega \mathcal{T}_{ps}(\omega) f_s(\omega) \tag{2.21}$$

$$> 0,$$

and

$$\lim_{\mu_p \to \infty} I^{(0)}(\mu_p, T_p) = \frac{1}{h} \int_{-\infty}^{\infty} d\omega \mathcal{T}_{ps}(\omega) \left[f_s(\omega) - \lim_{\mu_p \to \infty} f_p(\omega) \right]$$

$$= \frac{1}{h} \int_{-\infty}^{\infty} d\omega \mathcal{T}_{ps}(\omega) (f_s(\omega) - 1) \tag{2.22}$$

$$< 0.$$

This ensures at least one solution due to the continuity of the currents but does not ensure uniqueness.

We note that $I_p^{(0)}$ is monotonically decreasing along $\mathbf{dl} = (d\mu_p, 0)$

$$\Delta I_p^{(0)} = \int_{\mu_p}^{\mu_p'} \nabla I_p^{(0)} \cdot \mathbf{dl} = \int_{\mu_p}^{\mu_p'} -\mathcal{L}_{ps}^{(0)} d\mu_p \qquad (2.23)$$

due to the fact that $\mathcal{L}_{ps}^{(0)}$ is positive, and more explicitly

$$\begin{aligned}
\Delta I_p^{(0)} &= \frac{1}{h} \int_{-\infty}^{\infty} d\omega \mathcal{T}_{ps}(\omega)\left[f_p(\mu_p, T_p; \omega) - f_p(\mu_p', T_p; \omega)\right] \\
&> 0, \text{ if } \mu_p' < \mu_p \\
&< 0, \text{ if } \mu_p' > \mu_p.
\end{aligned} \qquad (2.24)$$

This implies the existence of a unique solution to $I_p^{(0)}(\mu_p, T_p) = 0$ for every $T_p \in (0, \infty)$ which we denote by $\mu_p = M(T_p)$, and Eq. (2.20) is implied by Eq. (2.24). ∎

We also note that the number current $[\mu_p = M(T_p)]$ contour is vertical when the temperature approaches absolute zero, as shown in Fig. 2.3, since $\mathcal{L}_{ps}^{(1)}/T_p \to 0$ as $T_p \to 0$, and implies a vanishing Seebeck coefficient for the probe-system junction near absolute zero.

An "ideal potentiometer" was initially proposed [19] by merely requiring $I_p^{(0)} = 0$. Subsequently, Büttiker [46, 47] clarified that this definition holds only near absolute zero due to the absence of thermoelectric corrections. Such a voltage probe determines the voltage uniquely at zero temperature in the linear response regime, and is relevant for experiments in mesoscopic circuits [48–51] which are carried out at cryogenic temperatures. However, at higher temperatures and/or larger bias voltages, where the sample may be heated by both the Joule and Peltier effects,

Fig. 2.3 Illustration of Lemma 2.1: The contour PQ shown in magenta cuts the number current contour $I_p^{(0)} = 0$ (or any $I_p^{(0)} = $ constant) exactly once. The contour line from P to Q is at a constant temperature ($T_p = $ constant) and illustrates the Clausius statement: The number current is monotonically decreasing along PQ. The system and bias conditions are detailed in Sect. 2.5.4

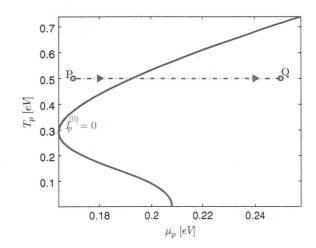

thermoelectric corrections to voltage measurements must be considered. Indeed, Bergfield and Stafford [40] argue that an *ideal voltage probe* must be required to equilibrate thermally with the system ($I_p^{(1)} = 0$) without which "a voltage will develop across the system-probe junction due to the *Seebeck effect*."

Voltage probes have been used extensively in the theoretical literature to mimic the effects of various scattering processes, such as inelastic scattering [46, 52–56] and dephasing [57–59] in mesoscopic systems. A modern variation of Büttiker's voltage probe, additionally requiring that the probe exchange no heat current, has been used to model inelastic scattering in quantum transport problems at finite temperature [21, 24, 60–62]. The probe technique, as a model for scattering, has also been extensively studied beyond the linear response regime [63–65].

Lemma 2.1 implies that a "voltage probe" (defined only by $I_p^{(0)} = 0$) requires the simultaneous specification of a probe temperature T_p so that $\mu_p = M(T_p)$ is uniquely determined. Figure 2.3 illustrates that the measured voltage shows a large dependence on the probe temperature. Therefore, it is important to define a simultaneous temperature measurement by imposing $I_p^{(1)}(\mu_p, T_p) = 0$.

Lemma 2.2 *The heat current contour defined by $I_p^{(1)}(\mu_p, T_p) = c$, where c is some constant, obeys the second law of thermodynamics:*

$$
\begin{aligned}
I_p^{(1)}(\mu_p, T_p') &> c, \ \ if \ T_p' < T_p \\
&< c, \ \ if \ T_p' > T_p.
\end{aligned}
\tag{2.25}
$$

Proof We follow an analogous argument to Lemma 2.1 and show the monotonicity of $I_p^{(1)}(\mu_p, T_p)$ along a certain contour in the μ_p-T_p plane. Naturally, the contour we choose is along a fixed μ_p [cf. Eq. (2.17)] since we know that $\mathcal{L}_{ps}^{(2)}$ is positive. Therefore we have $\Delta I_p^{(1)} = I_p^{(1)}(\mu_p, T_p') - I_p^{(1)}(\mu_p, T_p) = \int_{T_p}^{T_p'} \nabla I_p^{(1)}.\mathbf{dl}$, where $\mathbf{dl} = (0, dT_p)$ and explicitly,

$$
\begin{aligned}
\Delta I_p^{(1)} &= \frac{1}{h} \int_{-\infty}^{\infty} d\omega (\omega - \mu_p) \mathcal{T}_{ps}(\omega) \big[f_p(\mu_p, T_p; \omega) - f_p(\mu_p, T_p'; \omega) \big] \\
&> 0, \ \ if \ T_p' < T_p \\
&< 0, \ \ if \ T_p' > T_p.
\end{aligned}
\tag{2.26}
$$

This implies Eq. (2.25). ∎

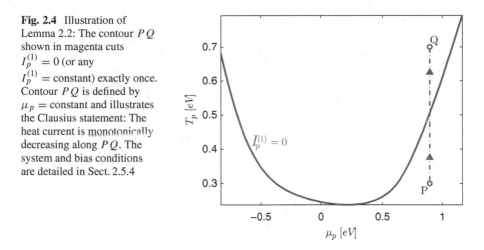

Fig. 2.4 Illustration of Lemma 2.2: The contour PQ shown in magenta cuts $I_p^{(1)} = 0$ (or any $I_p^{(1)} = $ constant) exactly once. Contour PQ is defined by $\mu_p = $ constant and illustrates the Clausius statement: The heat current is monotonically decreasing along PQ. The system and bias conditions are detailed in Sect. 2.5.4

We stated Lemma 2.2 with a constant c, not necessarily $c = 0$, unlike[4] Lemma 2.1. This is because we do *not a priori* know whether the contour $I_p^{(1)} = 0$ exists and we derive a necessary and sufficient condition for its existence in Sect. 2.5.

Analogous to Lemma 2.1, Lemma 2.2 implies that a "temperature probe" [19] (defined only by $I_p^{(1)} = 0$) requires the simultaneous specification of a probe voltage μ_p so that the temperature $T_p = \tau_0(\mu_p)$ is uniquely determined. Figure 2.4 illustrates that the measured temperature shows a large dependence on the probe voltage. Therefore, it becomes important to simultaneously measure the voltage by imposing $I_p^{(0)} = 0$. If the temperature probe is not allowed to equilibrate electrically with the system, a temperature difference will build up across the probe-system junction due to the *Peltier effect* thus leading to an error in the temperature measurement.

Clearly, depending upon the probe voltage, the "temperature probe" could measure any of a range of values thereby rendering the measurement somewhat meaningless (see Fig. 2.4). Analogously, the "voltage probe" could measure any of a range of values depending upon the probe temperature (see Fig. 2.3). *Thermoelectric probes* (also referred to as dual probes or voltage-temperature probes) treat temperature and voltage measurements on an equal footing and implicitly account for the thermoelectric corrections exactly. Only such a dual probe is in *both thermal and electrical equilibrium* with the system being measured, and therefore yields an

---

[4]Furthermore, the tangent vector [see Eq. (2.34)] along $I_p^{(1)} = c$ cannot be of magnitude zero since $\mathcal{L}_{ps}^{(2)}$ is strictly positive for $T_p \in (0, \infty)$. Therefore, the contour $I_p^{(1)} = c$ does not terminate for finite values of T_p and μ_p. This implies the existence of a function $\tau_c : (-\infty, \infty) \rightarrow (0, \infty)$ which defines

$$T_p = \tau_c(\mu_p)$$

for each point on $I_p^{(1)}(\mu_p, T_p) = c$.

unbiased measurement of both quantities. A mathematical proof of the uniqueness of a voltage and temperature measurement is therefore of fundamental importance.

We may also deduce that $T_p = 0$ cannot be obtained as a measurement outcome since

$$
\lim_{T_p \to 0} I_p^{(1)}(\mu_p, T_p) = \int_{-\infty}^{\infty} d\omega \, (\omega - \mu_p) \mathcal{T}_{ps}(\omega) \big[f_s(\omega)
$$

$$
- \lim_{T_p \to 0} f_p(\mu_p, T_p) \big]
$$

$$
= \int_{-\infty}^{\mu_p} d\omega \, (\omega - \mu_p) \mathcal{T}_{ps}(\omega)(f_s(\omega) - 1) \tag{2.27}
$$

$$
+ \int_{\mu_p}^{\infty} d\omega \, (\omega - \mu_p) \mathcal{T}_{ps}(\omega) f_s(\omega)
$$

$$
> 0,
$$

consistent with the third law of thermodynamics. However, temperatures arbitrarily close to absolute zero are, in principle, possible [9].

Lemmas 2.1 and 2.2 may be interpreted in terms of the Clausius statement of the second law [66]: "No process is possible whose sole effect is to transfer heat from a colder body to a warmer body." Lemma 2.2 gives us the direction in which heat will flow [see Eq. (2.26)] when the probe is biased away from the point of thermal equilibrium with the system, $I_p^{(1)}(\mu_p, T_p) = 0$: whenever the probe is *hotter* than the temperature corresponding to thermal equilibrium, with the chemical potential held constant, heat flows *out* of the probe and vice versa. Similarly, Lemma 2.1 gives us the direction in which particle flow occurs when the probe is biased away from the point of electrical equilibrium, $I_p^{(0)}(\mu_p, T_p) = 0$: whenever the probe is at a *higher* chemical potential than the one corresponding to electrical equilibrium, with temperature held constant, particles flow *out* of the probe and vice versa. Here we refer to electrical ($v = 0$, Lemma 2.1) and thermal ($v = 1$, Lemma 2.2) equilibration of the probe with the system under the local exchange of particles and energy. The system itself may be arbitrarily far from equilibrium and may possess no local equilibrium.

The problem of a unique measurement of a "voltage probe" (defined only by $I_p^{(0)} = 0$) or a "temperature probe" (defined only by $I_p^{(1)} = 0$) has been attempted previously by Jacquet and Pillet [6] for transport beyond linear response and, to our knowledge, is the only work in this direction. However, in Ref. [6], the bias conditions considered are quite restrictive and the result assumes noninteracting electrons. Lemmas 2.1 and 2.2, respectively, generalize the result to arbitrary bias conditions, and arbitrary interactions within a quantum electron system, while also providing a useful insight via the Clausius statement of the second law of thermodynamics. However, the question we would like to answer in this chapter pertains to the uniqueness of a *thermoelectric probe* measurement defined by both $I_p^{(0)} = 0$ *and* $I_p^{(1)} = 0$. A result for such dual probes has been obtained only in the linear response regime and for noninteracting electrons [21].

Theorem 2.1 *The coefficients $\mathcal{L}_{ps}^{(v)}$ satisfy the inequality*

$$\mathcal{L}_{ps}^{(0)}\mathcal{L}_{ps}^{(2)} - \left(\mathcal{L}_{ps}^{(1)}\right)^2 > 0. \tag{2.28}$$

Proof We may define functions $g(\omega)$ and $h(\omega)$ as

$$g(\omega) = \sqrt{\mathcal{T}_{ps}(\omega)\left(-\frac{\partial f_p}{\partial \omega}\right)} \tag{2.29}$$

and

$$h(\omega) = (\omega - \mu_p)\sqrt{\mathcal{T}_{ps}(\omega)\left(-\frac{\partial f_p}{\partial \omega}\right)}. \tag{2.30}$$

We note that $g(\omega)$ and $h(\omega)$ belong to $\mathbf{L}^2(\mathbb{R})$. Noting that g and h are real, we apply the Cauchy–Schwarz inequality

$$\left|\int_{-\infty}^{\infty} d\omega g(\omega)h(\omega)\right|^2 \leq \int_{-\infty}^{\infty} d\omega |g(\omega)|^2 \int_{-\infty}^{\infty} d\omega |h(\omega)|^2. \tag{2.31}$$

The integral appearing on the l.h.s. is $\mathcal{L}_{ps}^{(1)}$ while on the r.h.s. we have the product of $\mathcal{L}_{ps}^{(0)}$ and $\mathcal{L}_{ps}^{(2)}$, respectively. We drop the absolute value on the l.h.s. by noting that $\mathcal{L}_{ps}^{(1)}$ is real and write

$$\left(\mathcal{L}_{ps}^{(1)}\right)^2 \leq \mathcal{L}_{ps}^{(0)}\mathcal{L}_{ps}^{(2)}. \tag{2.32}$$

We drop the equality case above by noting that g and h are linearly independent except for the trivial case when $\mathcal{T}_{ps}(\omega) = 0 \;\forall \omega$ or when the probe coupling is narrowband $[\mathcal{T}_{ps}(\omega) = \bar{\gamma}\delta(\omega - \omega_0)]$. The latter scenario is discussed again in Sect. 2.5.3. ∎

The proof of Theorem 2.1 can be easily extended to show the positive-definiteness of the linear response matrices [44] widely used for elastic transport calculations[5] (e.g., in Refs. [45, 67]). Theorem 2.1 implies a positive thermal conductance (see e.g., Ref. [67]), which is necessary for positive entropy production consistent with the second law of thermodynamics. We therefore refer to this form as the Onsager's statement of the second law. Onsager, in his seminal work of 1931 on reciprocal relations [44, 68], asserted that the matrix of response coefficients must be positive since it would ensure positive entropy production consistent with the second law.

---

[5]Transport in a vast majority of mesoscopic and nanoscale conductors are dominated by elastic processes at room temperature. Theorem 2.1 proves the Onsager's phenomenological statement of the second law (1931) for quantum thermoelectric transport where elastic processes dominate the transport.

2.4.2 Uniqueness

Theorem 2.2 *The local temperature and voltage of a nonequilibrium quantum system, measured by a* thermoelectric probe, *is unique when it exists.*

Proof The tangent vectors $\mathbf{t}^{(\nu)}$ for $I_p^{(\nu)}$ are along

$$\mathbf{t}^{(0)} = \left(-\frac{\mathcal{L}_{ps}^{(1)}}{T_p}, \mathcal{L}_{ps}^{(0)} \right) \tag{2.33}$$

and

$$\mathbf{t}^{(1)} = \left(\frac{\mathcal{L}_{ps}^{(2)}}{T_p}, -\mathcal{L}_{ps}^{(1)} - I_p^{(0)} \right)$$

$$= \left(\frac{\mathcal{L}_{ps}^{(2)}}{T_p}, -\mathcal{L}_{ps}^{(1)} \right), \text{ if } I_p^{(0)} = 0, \tag{2.34}$$

respectively, such that we have

$$\int_{s_1}^{s_2} ds \frac{\mathbf{t}^{(\nu)} \cdot \nabla I_p^{(\nu)}}{|\mathbf{t}^{(\nu)}|} = 0, \tag{2.35}$$

where s is a scalar parameter that labels points along the contour $I_p^{(\nu)} = $ constant.

We now compute the change in $I_p^{(1)}$ along the contour $I_p^{(0)} = 0$. The points along $I_p^{(0)} = 0$ are labeled by the continuous parameter ξ such that $\mu_p = \mu_p(\xi)$ and $T_p = T_p(\xi)$. ξ is chosen to be increasing with increasing temperature. The change $\Delta I_p^{(1)}$ becomes

$$\Delta I_p^{(1)} = \int_{\xi_1}^{\xi_2} d\xi \frac{\mathbf{t}^{(0)} \cdot \nabla I_p^{(1)}}{|\mathbf{t}^{(0)}|}$$

$$= \int_{\xi_1}^{\xi_2} d\xi \frac{1}{|\mathbf{t}^{(0)}| T_p} \left((\mathcal{L}_{ps}^{(1)})^2 - \mathcal{L}_{ps}^{(0)} \mathcal{L}_{ps}^{(2)} \right) \tag{2.36}$$

$$> 0 \text{ if } \xi_2 < \xi_1$$

$$< 0 \text{ if } \xi_2 > \xi_1,$$

due to Theorem 2.1. Therefore $I_p^{(1)} = 0$ (or for that matter $I_p^{(1)} = c$, for any c) is satisfied at most at a single point along $I_p^{(0)} = 0$. ∎

Theorem 2.1 is a form of the second law of thermodynamics that gives us the direction in which the heat current flows along the contour $I_p^{(0)} = 0$ [see Eq. (2.36)].

The heat current $I_p^{(1)}$ decreases monotonically along the contour $I_p^{(0)} = 0$. Therefore we may find only one point along $I_p^{(0)} = 0$ that also satisfies $I_p^{(1)} = 0$, which implies a unique solution to Eq. (2.13) when it exists.

As mentioned previously, Onsager pointed out [44, 68] that for positive entropy production, the linear response matrix will have to be positive-definite (which translates to our condition in Theorem 2.1). However, that analysis rests upon the assumption of linear response near equilibrium. Our result in Theorem 2.1 does not require such a condition for the nonequilibrium state of the system but instead emerges out of the analysis of the currents flowing into a weakly coupled probe. In addition, we obtain a mathematical proof of Onsager's phenomenological statement for the case of quantum thermoelectric transport. We point out that Theorem 2.1 holds even when the physically expected postulate 2.1 fails, making the uniqueness result in Theorem 2.2 very general.

2.5 Existence

A unique local measurement of temperature and voltage is only part of our main problem. An equally important part is to derive the conditions for the existence of a solution. The main idea behind this analysis is to follow the number current contour $I_p^{(0)} = 0$ and ask what happens to the heat current $I_p^{(1)}$ as we traverse towards higher and higher temperatures $T_p \rightarrow \infty$. We noted that near $T_p = 0$ the heat current into the probe must be positive, consistent with the third law of thermodynamics [see Eq. (2.27)]. Since we know that the heat current is monotonically decreasing along the number current contour (Theorem 2.2), we could guess whether or not a solution occurs depending upon the asymptotic value of the heat current along that contour as $T_p \rightarrow \infty$. In this way, we find a necessary and sufficient condition for the existence of a solution while analyzing the problem for positive temperatures (see Fig. 2.5 for an illustration of this case). On the other hand, when this condition is not met, one can immediately prove that a negative temperature must satisfy the measurement condition $I_p^{(\nu)} = 0$, $\nu = \{0, 1\}$. This latter condition corresponds to a system exhibiting local population inversion which leads to negative temperature [69] solutions, as illustrated in Fig. 2.6.

Our results here are again completely general and are valid for electron systems with arbitrary interactions, arbitrary steady-state bias conditions, and for any weakly coupled probe. However, our analysis here leads us to demarcate between two extremes of the probe-system coupling. We conclude that an *ideal probe* is one which operates in the *broadband limit*. A measurement by such a probe depends only on the properties of the system that it couples to and is independent of the spectral properties of the probe itself. The broadband limit lends itself to an easier physical interpretation of the population inversion condition as well and we discuss this important limit in Sect. 2.5.2. The other extreme is that of a *narrowband probe* which is capable of probing the system at just one value of energy, leading to a

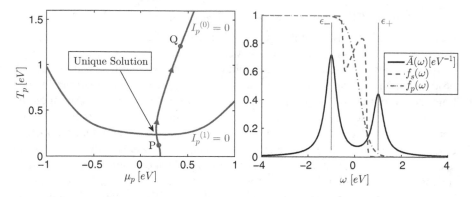

Fig. 2.5 Left panel: Illustration of Theorem 2.2 for positive temperatures. The contour PQ along $I_p^{(0)} = 0$ (shown in blue) cuts the contour $I_p^{(1)} = 0$ (shown in red) exactly once. Contour PQ illustrates a certain statement of the second law of thermodynamics: The heat current is monotonically decreasing along PQ (thus implying uniqueness). Right panel: The local spectrum sampled by the probe $\bar{A}(\omega)$ (black), the nonequilibrium distribution function $f_s(\omega)$ (red), and the probe Fermi–Dirac distribution $f_p(\omega)$ (blue) corresponding to the unique solution in the left panel. The resonances in the spectrum $\bar{A}(\omega)$ correspond to the eigenstates of the closed two-level Hamiltonian (see Sect. 2.5.4) $\epsilon_\pm = \pm 1$ shown in magenta. The Fermi–Dirac distribution is monotonically decreasing with energy, and corresponds to a situation with positive temperature (no net population inversion). The necessary and sufficient condition for the existence of a positive temperature solution is stated in Theorem 2.3

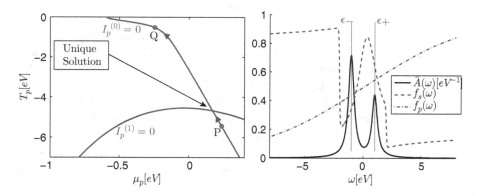

Fig. 2.6 Left panel: Illustration of Theorem 2.2 for negative temperatures. The contour PQ along $I_p^{(0)} = 0$ (shown in blue) cuts the contour $I_p^{(1)} = 0$ (shown in red) exactly once. Contour PQ illustrates a certain statement of the second law of thermodynamics: The heat current is monotonically decreasing along PQ (thus implying uniqueness). Right panel: The local spectrum sampled by the probe $\bar{A}(\omega)$ (black, and nearly unchanged from Fig. 2.5), the nonequilibrium distribution function $f_s(\omega)$ (red), and the probe Fermi–Dirac distribution $f_p(\omega)$ (blue) which corresponds to the unique solution (shown in the left panel). The resonances in the spectrum $\bar{A}(\omega)$ correspond to the eigenstates of the closed two-level Hamiltonian (see Sect. 2.5.4) $\epsilon_\pm = \pm 1$ shown in magenta. The system has a net population inversion, satisfying the conditions of Corollary 2.3.1, and the probe Fermi–Dirac distribution is monotonically increasing with energy, corresponding to a negative temperature

nonunique measurement (see also the proof of Theorem 2.1), and is discussed in
Sect. 2.5.3. Only this pathological case leads to an exception to Theorem 2.2.

The simplest system which could, in principle, exhibit population inversion is a
two-level system. Therefore, our results, including that of the previous section, have
been illustrated by using a two-level system. The details of the nonequilibrium two-
level system and its coupling to the thermoelectric probe are given in Sect. 2.5.4.

Our analysis starts with a rearrangement of the currents given by Eq. (2.2) and
a restatement of the measurement condition [see Eq. (2.13)] in terms of energy
currents and we also define some useful quantities along the way. We may rewrite
the number current in Eq. (2.2) as

$$I_p^{(0)} = \langle \dot{N} \rangle |_{f_s} - \langle \dot{N} \rangle |_{f_p} \tag{2.37}$$

where

$$\langle \dot{N} \rangle |_{f_s} \equiv \frac{1}{h} \int_{-\infty}^{\infty} d\omega \, \mathcal{T}_{ps}(\omega) f_s(\omega), \tag{2.38}$$

and similarly

$$\langle \dot{N} \rangle |_{f_p} \equiv \frac{1}{h} \int_{-\infty}^{\infty} d\omega \, \mathcal{T}_{ps}(\omega) f_p(\omega). \tag{2.39}$$

The quantity $\langle \dot{N} \rangle |_{f_s}$ is the rate of particle flow into the probe from the system while
$\langle \dot{N} \rangle |_{f_p}$ gives the rate of particle flow out of the probe and into the system.

Similarly, the rate of energy flow into the probe from the system is

$$\langle \dot{E} \rangle |_{f_s} \equiv \frac{1}{h} \int_{-\infty}^{\infty} d\omega \, \omega \mathcal{T}_{ps}(\omega) f_s(\omega), \tag{2.40}$$

while

$$\langle \dot{E} \rangle |_{f_p} \equiv \frac{1}{h} \int_{-\infty}^{\infty} d\omega \, \omega \mathcal{T}_{ps}(\omega) f_p(\omega) \tag{2.41}$$

gives the rate of energy outflux from the probe back into the system. The net energy
current flowing into the probe is given by $I_p^E = \langle \dot{E} \rangle |_{f_s} - \langle \dot{E} \rangle |_{f_p}$.

The local equilibration conditions in Eq. (2.13) now become

$$\langle \dot{N} \rangle |_{f_p} = \langle \dot{N} \rangle |_{f_s}$$
$$\langle \dot{E} \rangle |_{f_p} = \langle \dot{E} \rangle |_{f_s}. \tag{2.42}$$

The equation for the rate of energy flow above is equivalent to the condition $I_p^{(1)} = 0$
when $I_p^{(0)} = 0$ since

$$I_p^E(\mu_p, T_p) \equiv \langle \dot{E} \rangle|_{f_s} - \langle \dot{E} \rangle|_{f_p} = I_p^{(1)} + \mu_p I_p^{(0)}. \tag{2.43}$$

The l.h.s. in Eq. (2.42) depends upon the probe parameters (temperature and voltage) while the r.h.s. is fixed for a given nonequilibrium system with a given local distribution function $f_s(\omega)$. The probe measures the appropriate voltage and temperature when it exchanges no net charge and energy with the system.

We may introduce a characteristic rate of particle flow [cf. Eq. (2.9)] as

$$\begin{aligned} \langle \dot{N} \rangle|_{f\equiv 1} &= \frac{1}{h} \int_{-\infty}^{\infty} d\omega \mathcal{T}_{ps}(\omega) \\ &\equiv \frac{\gamma_p}{\hbar}. \end{aligned} \tag{2.44}$$

This leads to the following inequalities:

$$\begin{aligned} 0 &< \langle \dot{N} \rangle|_{f_s} < \frac{\gamma_p}{\hbar}, \\ 0 &< \langle \dot{N} \rangle|_{f_p} < \frac{\gamma_p}{\hbar}. \end{aligned} \tag{2.45}$$

The l.h.s. in the inequality for $\langle \dot{N} \rangle|_{f_s}$ above excludes $f_s(\omega) \equiv 0$ while the r.h.s. excludes $f_s(\omega) = 1 \; \forall \omega \in \mathbb{R}$, and we retain the strict inequalities imposed by Eq. (2.45) (see also Eqs. (2.11) and (2.12) and the preceding discussion).

We similarly introduce a characteristic rate for the energy flow between the system and probe:

$$\begin{aligned} \langle \dot{E} \rangle|_{f\equiv 1} &= \frac{1}{h} \int_{-\infty}^{\infty} d\omega \, \omega \mathcal{T}_{ps}(\omega) \\ &\equiv \frac{\gamma_p}{\hbar} \omega_c, \end{aligned} \tag{2.46}$$

where $\omega_c < \infty$ (due to postulate 2.1) can be interpreted as the centroid of the probe-sample transmission function. We find that $\omega_c \to \infty$ necessarily implies a positive temperature solution. We remind the reader that $\omega_c \to -\infty$ is physically impossible due to the principle that any physical system must have a lower bound for the energy ($\langle H \rangle \geq -c$ for some finite $c \in \mathbb{R}$).

The quantities $\langle \dot{N} \rangle|_{f_s}$, $\langle \dot{N} \rangle|_{f_p}$, $\langle \dot{N} \rangle|_{f\equiv 1}$, $\langle \dot{E} \rangle|_{f_s}$, $\langle \dot{E} \rangle|_{f_p}$, $\langle \dot{E} \rangle|_{f\equiv 1}$ are all finite due to postulate 2.1 [see Eqs. (2.9–2.12)].

2.5.1 Asymptotic Properties and Conditions for the Existence of a Solution

Traversing along $I_p^{(0)} = 0$ results in a monotonically *decreasing* heat current $I_p^{(1)}$ (Theorem 2.2). Here, we traverse the contour from low temperatures ($T_p \to 0$) to high temperatures ($T_p \to \infty$) as discussed in Theorem 2.2. This implies a monotonically *increasing* $\langle \dot{F} \rangle|_{f_p}$ due to Eq. (2.43). We proceed to calculate the asymptotic value of $\langle \dot{E} \rangle|_{f_p}$ along the number current contour.

Let the asymptotic scaling of $\mu_p = M(T_p)$ defined by the contour $I_p^{(0)}(\mu_p, T_p) = 0$ (Lemma 2.1) be

$$\lim_{T_p \to \infty} \frac{M(T_p)}{T_p} = \Lambda. \tag{2.47}$$

We use the above limiting value to calculate $\langle \dot{N} \rangle|_{f_p}$ along the contour $\mu_p = M(T_p)$:

$$\lim_{T_p \to \infty} \langle \dot{N} \rangle|_{f_p} = \frac{1}{h} \int_{-\infty}^{\infty} d\omega \mathcal{T}_{ps}(\omega) \times \lim_{T_p \to \infty} \frac{1}{1 + \exp\left\{ \left(\frac{\omega - M(T_p)}{T_p} \right) \right\}}$$

$$= \frac{1}{h} \int_{-\infty}^{\infty} d\omega \mathcal{T}_{ps}(\omega) \frac{1}{1 + \exp\{(-\Lambda)\}} \tag{2.48}$$

$$= \frac{1}{1 + \exp\{(-\Lambda)\}} \frac{\gamma_p}{\hbar}.$$

The above limiting value satisfies the inequality in Eq. (2.45) for any $\Lambda \in \mathbb{R}$. The points on the contour satisfy $\langle \dot{N} \rangle|_{f_p} = \langle \dot{N} \rangle|_{f_s}$ by construction; therefore, Λ is computed from the equation

$$\frac{1}{1 + \exp\{(-\Lambda)\}} \frac{\gamma_p}{\hbar} = \langle \dot{N} \rangle|_{f_s}. \tag{2.49}$$

It is important to note that the asymptotic scaling defined by Eq. (2.47) does not mean that the scaling is linear. For example, a sublinear scaling $M(T_p) = \alpha T_p^n$ with $n < 1$ merely corresponds to $\Lambda = 0$ which could satisfy Eq. (2.49) if the nonequilibrium system is prepared in that way. However, $\Lambda \to \pm\infty$ do not obey the strict inequality in Eq. (2.45). $\Lambda \to \infty$ corresponds to a trivial and unphysical nonequilibrium distribution $f_s(\omega) \equiv 1$, and likewise, $\Lambda \to -\infty$ corresponds to $f_s(\omega) \equiv 0 \ \forall \omega$.

The asymptotic value of $\langle \dot{E} \rangle|_{f_p}$ along the $I_p^{(0)} = 0$ contour is simply

$$\lim_{T_p \to \infty} \langle \dot{E} \rangle|_{f_p} = \frac{1}{h} \int_{-\infty}^{\infty} d\omega\, \omega \mathcal{T}_{ps}(\omega) \times \lim_{T_p \to \infty} \frac{1}{1 + \exp\left(\frac{\omega - M(T_p)}{T_p}\right)}$$

$$= \frac{1}{h} \int_{-\infty}^{\infty} d\omega\, \omega \mathcal{T}_{ps}(\omega) \frac{1}{1 + \exp\{(-\Lambda)\}} \tag{2.50}$$

$$= \frac{1}{1 + \exp\{(-\Lambda)\}} \frac{\gamma_p}{\hbar} \omega_c$$

$$= \omega_c \langle \dot{N} \rangle|_{f_s}.$$

Theorem 2.3 *A positive temperature solution exists if and only if there is no net population inversion, i.e., when*

$$\frac{\langle \dot{E} \rangle|_{f_s}}{\langle \dot{N} \rangle|_{f_s}} < \omega_c. \tag{2.51}$$

Proof $\langle \dot{E} \rangle|_{f_p}/\langle \dot{N} \rangle|_{f_s} < \langle \dot{E} \rangle|_{f_s}/\langle \dot{N} \rangle|_{f_s}$ when $T_p \to 0$ along the contour $I_p^{(0)} = 0$ [see Eqs. (2.27) and (2.43)]. The asymptotic limit of $\langle \dot{E} \rangle|_{f_p}/\langle \dot{N} \rangle|_{f_s}$ is ω_c [see Eq. (2.50)]. $\langle \dot{E} \rangle|_{f_p}$ is continuous $\forall\, \mu_p \in (-\infty, \infty)$, $T_p \in (0, \infty)$ and is monotonically increasing along $I_p^{(0)} = 0$ (Theorem 2.2). We use the intermediate value theorem.   ∎

Corollary 2.3.1 *There exists a negative temperature solution for a nonequilibrium system with net population inversion, i.e., when*

$$\frac{\langle \dot{E} \rangle|_{f_s}}{\langle \dot{N} \rangle|_{f_s}} > \omega_c. \tag{2.52}$$

Proof Let $f_p(\mu_p, T_p)$ be the Fermi–Dirac distribution with $T_p > 0$; we define the Fermi–Dirac distribution $f_p^- \equiv f_p(\mu_p, -T_p) = 1 - f_p$.

$$I_p^{(\nu)}(\mu_p, -T_p) = \frac{1}{h} \int_{-\infty}^{\infty} d\omega (\omega - \mu_p)^\nu \mathcal{T}_{ps}(\omega) \left[f_s(\omega) - (1 - f_p(\omega)) \right]$$

$$= \frac{1}{h} \int_{-\infty}^{\infty} d\omega (\omega - \mu_p)^\nu \mathcal{T}_{ps}(\omega) \left[f_p(\omega) - (1 - f_s(\omega)) \right]$$

$$= \frac{1}{h} \int_{-\infty}^{\infty} d\omega (\omega - \mu_p)^\nu \mathcal{T}_{ps}(\omega) \left[f_p(\omega) - f_s^-(\omega) \right]$$

$$\equiv -I_p^{(\nu)-} \tag{2.53}$$

$I_p^{(\nu)-} = 0$ with $\nu = \{0, 1\}$ is now understood to solve the complementary nonequilibrium system with $f_s^-(\omega) \equiv 1 - f_s(\omega)$.

$f_s^-(\omega)$ is of course a completely valid nonequilibrium distribution function and satisfies Eq. (2.8). We apply Theorem 2.3 and find that

$$\langle \dot{E} \rangle|_{f_s^-} < \omega_c \langle \dot{N} \rangle|_{f_s^-}$$

$$\frac{\gamma_p}{\hbar} \omega_c - \langle \dot{E} \rangle|_{f_s} < \omega_c \left(\frac{\gamma_p}{\hbar} - \langle \dot{N} \rangle|_{f_s} \right)$$

$$-\langle \dot{E} \rangle|_{f_s} < -\omega_c \langle \dot{N} \rangle|_{f_s}$$

$$\langle \dot{E} \rangle|_{f_s} > \omega_c \langle \dot{N} \rangle|_{f_s}.$$

(2.54)

For the case that $\langle \dot{E} \rangle|_{f_s} = \omega_c \langle \dot{N} \rangle|_{f_s}$, $T_p = \pm\infty$, corresponding to $f_p = 1/2$, independent of energy. ∎

2.5.2   Ideal Probes: The Broadband Limit

In the broadband limit the probe-system coupling becomes energy independent and we may write $\Gamma^p(\omega) = \Gamma^p(\mu_0)$. The spectrum of the system, sampled locally by the probe, is given by

$$\bar{A}(\omega) \equiv \frac{\mathrm{Tr}\{\Gamma^p(\omega)A(\omega)\}}{\mathrm{Tr}\{\Gamma^p(\omega)\}}$$

$$= \frac{\mathrm{Tr}\{\Gamma^p(\mu_0)A(\omega)\}}{\mathrm{Tr}\{\Gamma^p(\mu_0)\}}.$$

(2.55)

The occupancy and energy of the system, respectively, are given by

$$\langle N \rangle|_{f_s} = \int_{-\infty}^{\infty} d\omega \bar{A}(\omega) f_s(\omega)$$

$$\langle E \rangle|_{f_s} = \int_{-\infty}^{\infty} d\omega \, \omega \bar{A}(\omega) f_s(\omega).$$

(2.56)

The measurement conditions in Eq. (2.13) become simply [7]

$$\langle N \rangle|_{f_p} = \langle N \rangle|_{f_s}$$

$$\langle E \rangle|_{f_p} = \langle E \rangle|_{f_s}.$$

(2.57)

The above equations imply that an *ideal measurement* of voltage and temperature constitutes a measurement of the zeroth and first moments of the local energy distribution of the system. That is to say, when the probe is in local equilibrium with the nonequilibrium system, the local occupancy and energy of the system are the same as they would be if the system's local spectrum were populated by the equilibrium Fermi–Dirac distribution $f_p \equiv f_p(\mu_p, T_p)$ of the probe.

We may now write the condition for the existence of a positive temperature solution (Theorem 2.3) simply as

$$\frac{\langle E \rangle|_{f_s}}{\langle N \rangle|_{f_s}} < \omega_c, \tag{2.58}$$

where ω_c is the centroid of the spectrum given by

$$\omega_c = \int_{-\infty}^{\infty} d\omega\, \omega \bar{A}(\omega). \tag{2.59}$$

The condition in Eq. (2.58) implies the following: Given some nonequilibrium distribution function f_s, one can have a positive temperature solution if and only if the average energy per particle is smaller than the centroid of the spectrum. In other words, a positive temperature solution exists if and only if there is no net population inversion. Similarly, Corollary 2.3.1 states that there exists a negative temperature solution for a system exhibiting population inversion:

$$\frac{\langle E \rangle|_{f_s}}{\langle N \rangle|_{f_s}} > \omega_c. \tag{2.60}$$

The advantage of the broadband limit is that one may write the measurement conditions, as well as the condition for the existence of a solution, in terms of the local expectation values of the energy and occupancy directly, instead of using the rate of particle and energy flow into the probe. We also do not need to introduce a "characteristic tunneling rate." We note that ω_c in Eq. (2.59) is the centroid since the local spectrum $\bar{A}$ normalizes to unity within the broadband limit (see Appendix A, Sect. A.1).

A local measurement by a weakly coupled broadband thermoelectric probe is *ideal* in the sense that the result is independent of the properties of the probe, and depends only on the nonequilibrium state of the system and the subsystem thereof sampled by the probe. Such a measurement provides more than just an operational definition of the local temperature and voltage of a nonequilibrium quantum system, since the thermodynamic variables are determined directly by the moments (2.56) of the local (nonequilibrium) energy distribution.

2.5.3   Pathological Probes: The Narrowband Limit

A narrowband probe is one that samples the system only within a very narrow window of energy. The extreme case of such a probe-system coupling would be a Dirac-delta function:

$$\Gamma_p(\omega) - 2\pi\, V_p^\dagger V_p \delta(\omega - \omega_0), \tag{2.61}$$

which gives $\mathcal{T}_{ps}(\omega) = 2\pi\, \mathrm{Tr}\left\{V_p A(\omega) V_p^\dagger\right\}\delta(\omega - \omega_0)$ which we write simply as

$$\mathcal{T}_{ps}(\omega) = \gamma(\omega)\,\delta(\omega - \omega_0), \tag{2.62}$$

where $\gamma(\omega) = 2\pi\, \mathrm{Tr}\left\{V_p A(\omega) V_p^\dagger\right\}$ has dimensions of energy.

We previously noted that Theorem 2.1 does not hold for $\mathcal{T}_{ps}$ given by Eq. (2.62). One can verify straightforwardly that, for a probe-sample transmission that is extremely narrow, we will have

$$\mathcal{L}_{ps}^{(0)}\mathcal{L}_{ps}^{(2)} - \left(\mathcal{L}_{ps}^{(1)}\right)^2 = 0. \tag{2.63}$$

This results in a nonunique solution since following the proof of Theorem 2.2 would give us [see Eq. (2.36)] $\Delta I_p^{(1)} = 0$. In fact, it would lead to a family of solutions.

We may solve for the solution explicitly. The number current reduces to

$$I_p^{(0)} = \frac{\gamma(\omega_0)}{h}\left(f_p(\omega_0) - f_s(\omega_0)\right), \tag{2.64}$$

while the heat current is given by

$$I_p^{(1)} = (\omega_0 - \mu_p)\frac{\gamma(\omega_0)}{h}\left(f_p(\omega_0) - f_s(\omega_0)\right), \tag{2.65}$$

which trivially vanishes for vanishing number current. Therefore, the family of solutions to the measurement is simply given by

$$f_p(\omega_0; \mu_p, T_p) = f_s(\omega_0), \tag{2.66}$$

which is linear in the $\mu_p - T_p$ plane and is given by

$$\mu_p = \omega_0 - T_p \log\left(\frac{1 - f_s(\omega_0)}{f_s(\omega_0)}\right). \tag{2.67}$$

$f_s(\omega)$ has the following explicit form:

$$f_s(\omega) = \frac{\text{Tr}\left\{V_p G^<(\omega) V_p^\dagger\right\}}{2\pi i \,\text{Tr}\left\{V_p A(\omega) V_p^\dagger\right\}}. \tag{2.68}$$

A *narrowband probe* is therefore unsuitable for thermoelectric measurements. Even if a probe were to sample the system at just two distinct energies ω_1 and ω_2, Theorem 2.1 would hold and the thermoelectric measurement would be unique. Indeed, the narrowband probe is a pathological case whose only function is to highlight a certain theoretical limitation for the measurement of the temperature and voltage.

2.5.4   Example: Two-Level System

Net population inversion is essentially a quantum phenomenon, since classical Hamiltonians are generally unbounded above due to the kinetic energy term, i.e., there does not exist a finite $c \in \mathbb{R}$ that satisfies $\langle H \rangle < c$. In other words, $\omega_c \to \infty$ generally holds for classical systems and negative temperatures are not possible. The simplest quantum system where a net population inversion can be achieved is a two-level system (depicted in Fig. 2.7). We therefore illustrated our results for a two-level system in Figs. 2.3, 2.4, 2.5 and 2.6.

The system Hamiltonian here was taken to be

$$H = \begin{bmatrix} \epsilon_1 & V \\ V^* & \epsilon_2 \end{bmatrix}, \tag{2.69}$$

whose values were set as $V = \frac{2(1-i)}{3}$, $\epsilon_1 = 1/3$ and $\epsilon_2 = -1/3$, such that the eigenvalues are $\epsilon_\pm = \pm 1$ and units could be read to be in eV. We introduce two reservoirs that are strongly coupled locally to each site with $\Gamma_1 = \text{diag}(0.5, 0)$ and $\Gamma_2 = \text{diag}(0, 0.5)$, while the probe coupling is taken as $\Gamma_p = \text{diag}(0.01, 0.1)$, which is about five times weaker than the coupling to the reservoirs that drive the system.

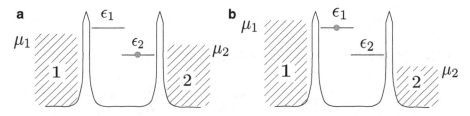

Fig. 2.7 Schematic diagram of a two-level system coupled to two electron reservoirs under a voltage bias. (**a**) Bias condition not leading to population inversion. (**b**) Bias condition leading to population inversion due to direct injection into excited state

We used two different bias conditions: (a) To illustrate the case without a net population inversion in Figs. 2.3, 2.4 and 2.5, the reservoirs had a symmetric ($\mu_1 + \mu_2 = 0$) voltage bias $\mu_1 - \mu_2 = 1$ eV; (b) to illustrate the case with a net population inversion in Fig. 2.6, the reservoirs had a symmetric voltage bias of $\mu_1 - \mu_2 = 4$ eV. The two reservoirs are held at $T = 300$ K for both cases.

It has been previously noted that the probe-system coupling strength does not strongly affect the measured temperature and voltage even when varied over several orders of magnitude [35], but we remind the reader that our theoretical results depend upon the assumption of a weakly coupled probe (noninvasive measurements). How *weak* is *weak enough* is a more subtle theoretical question. We present a detailed analysis of this point in Appendix B on noninvasive measurements. Numerically, however, we do find that the probe measurements are not much altered even when the probe coupling strength is comparable to that of the strongly coupled reservoirs.

2.6 Conclusions

The local temperature and voltage of a nonequilibrium quantum system are defined in terms of the equilibration of a noninvasive thermoelectric probe, locally coupled to the system. The simultaneous temperature and voltage measurement is shown to be unique for any system of fermions in steady state, arbitrarily far from equilibrium, with arbitrary interactions within the system, and the conditions for the existence of a solution are derived. In particular, it is shown that a positive temperature solution exists provided the system does not have a net local population inversion; in the case of population inversion, a unique negative temperature solution is shown to exist. These results provide a firm mathematical foundation for temperature and voltage measurements in quantum systems far from equilibrium.

Our analysis reveals that a simultaneous temperature and voltage measurement is uniquely determined by the local spectrum and nonequilibrium distribution of the system [cf. Eq. (2.42)], and is independent of the properties of the probe for broadband coupling (ideal probe). Such a measurement therefore provides a *fundamental definition* of local temperature and voltage which is not merely operational.

In contrast, prior theoretical work relied almost exclusively on operational definitions [6, 19–28] leading to a competing panoply of often contradictory predictions for the measurement of such basic observables as temperature and voltage. Measurements of temperature or voltage, taken separately (see, e.g., Refs. [6, 19]), are shown to be ill-posed: a thermometer out of electrical equilibrium with a system produces an error due to the Peltier effect across the probe-sample junction, while a potentiometer out of thermal equilibrium with a system produces an error due to the Seebeck effect. Most importantly, such measurements are not unique even for the same probe and system: the probe's temperature influences the voltage measurement and vice versa.

Our results put the local thermodynamic variables temperature and voltage on a mathematically rigorous footing for fermion systems under very general nonequilibrium steady-state conditions—a necessary first step toward the construction of nonequilibrium thermodynamics [2–9]. Our analysis includes the effect of interactions with bosonic degrees of freedom (e.g., photons, phonons, etc.) on the fermions. However, the temperatures of the bosons themselves [30, 31] were not addressed in the present analysis. Moreover, we did not explicitly consider magnetic systems, which require separate consideration of the spin degree of freedom and its polarization. Future investigation of probes that exchange bosonic or spin excitations may enable similarly rigorous analyses of local thermodynamic variables in bosonic and magnetic systems, respectively.

References

1. A.S. Eddington, *The Nature of the Physical World* (The University Press, Cambridge, 1929)
2. D. Ruelle, J. Stat. Phys. **98**(1), 57 (2000). https://doi.org/10.1023/A:1018618704438. http://dx.doi.org/10.1023/A:1018618704438
3. J. Casas-Vázquez, D. Jou, Rep. Prog. Phys. **66**(11), 1937 (2003). http://stacks.iop.org/0034-4885/66/i=11/a=R03
4. G. Lebon, D. Jou, *Understanding Non-equilibrium Thermodynamics* (Springer, Berlin, 2008). https://doi.org/10.1007/978-3-540-74252-4
5. L.F. Cugliandolo, J. Phys. A Math. Theor. **44**(48), 483001 (2011). http://stacks.iop.org/1751-8121/44/i=48/a=483001
6. P.A. Jacquet, C.A. Pillet, Phys. Rev. B **85**, 125120 (2012). https://doi.org/10.1103/PhysRevB.85.125120. http://link.aps.org/doi/10.1103/PhysRevB.85.125120
7. C.A. Stafford, Phys. Rev. B **93**, 245403 (2016). https://doi.org/10.1103/PhysRevB.93.245403. http://link.aps.org/doi/10.1103/PhysRevB.93.245403
8. M. Esposito, M.A. Ochoa, M. Galperin, Phys. Rev. Lett. **114**, 080602 (2015). https://doi.org/10.1103/PhysRevLett.114.080602. http://link.aps.org/doi/10.1103/PhysRevLett.114.080602
9. A. Shastry, C.A. Stafford, Phys. Rev. B **92**, 245417 (2015). https://doi.org/10.1103/PhysRevB.92.245417. http://link.aps.org/doi/10.1103/PhysRevB.92.245417
10. P. Muralt, D.W. Pohl, Appl. Phys. Lett. **48**(8), 514 (1986). http://dx.doi.org/10.1063/1.96491. http://scitation.aip.org/content/aip/journal/apl/48/8/10.1063/1.96491
11. T. Kanagawa, R. Hobara, I. Matsuda, T. Tanikawa, A. Natori, S. Hasegawa, Phys. Rev. Lett. **91**, 036805 (2003). https://doi.org/10.1103/PhysRevLett.91.036805. http://link.aps.org/doi/10.1103/PhysRevLett.91.036805
12. A. Bannani, C.A. Bobisch, R. Moeller, Rev. Sci. Instrum. **79**(8), 083704 (2008). https://doi.org/10.1063/1.2968111
13. F. Luepke, S. Korte, V. Cherepanov, B. Voigtlaender, Rev. Sci. Instrum. **86**(12), 123701 (2015). https://doi.org/10.1063/1.4936079
14. C.C. Williams, H.K. Wickramasinghe, Appl. Phys. Lett. **49**(23), 1587 (1986). http://dx.doi.org/10.1063/1.97288. http://scitation.aip.org/content/aip/journal/apl/49/23/10.1063/1.97288
15. K. Kim, J. Chung, G. Hwang, O. Kwon, J.S. Lee, ACS Nano **5**(11), 8700 (2011). https://doi.org/10.1021/nn2026325. http://dx.doi.org/10.1021/nn2026325. PMID: 21999681
16. Y.J. Yu, M.Y. Han, S. Berciaud, A.B. Georgescu, T.F. Heinz, L.E. Brus, K.S. Kim, P. Kim, Appl. Phys. Lett. **99**(18), 183105 (2011). http://dx.doi.org/10.1063/1.3657515. http://scitation.aip.org/content/aip/journal/apl/99/18/10.1063/1.3657515
17. K. Kim, W. Jeong, W. Lee, P. Reddy, ACS Nano **6**(5), 4248 (2012). https://doi.org/10.1021/nn300774n. http://dx.doi.org/10.1021/nn300774n. PMID: 22530657

18. F. Menges, H. Riel, A. Stemmer, B. Gotsmann, Nano Lett. **12**(2), 596 (2012). https://doi.org/10.1021/nl203169t. http://dx.doi.org/10.1021/nl203169t. PMID: 22214277
19. H.L. Engquist, P.W. Anderson, Phys. Rev. B **24**, 1151 (1981). https://doi.org/10.1103/PhysRevB.24.1151. http://link.aps.org/doi/10.1103/PhysRevB.24.1151
20. Y. Dubi, M. Di Ventra, Nano Lett. **9**, 97 (2009)
21. P.A. Jacquet, J. Stat. Phys. **134**(4), 709 (2009). https://doi.org/10.1007/s10955-009-9697-1. http://dx.doi.org/10.1007/s10955-009-9697-1
22. Y. Dubi, M. Di Ventra, Phys. Rev. E **79**, 042101 (2009). https://doi.org/10.1103/PhysRevE.79.042101. http://link.aps.org/doi/10.1103/PhysRevE.79.042101
23. A. Caso, L. Arrachea, G.S. Lozano, Phys. Rev. B **83**, 165419 (2011). https://doi.org/10.1103/PhysRevB.83.165419. http://link.aps.org/doi/10.1103/PhysRevB.83.165419
24. D. Sánchez, L. Serra, Phys. Rev. B **84**, 201307 (2011). https://doi.org/10.1103/PhysRevB.84.201307. http://link.aps.org/doi/10.1103/PhysRevB.84.201307
25. A. Caso, L. Arrachea, G.S. Lozano, Phys. Rev. B **81**(4), 041301 (2010). https://doi.org/10.1103/PhysRevB.81.041301
26. J.P. Bergfield, S.M. Story, R.C. Stafford, C.A. Stafford, ACS Nano **7**(5), 4429 (2013). https://doi.org/10.1021/nn401027u
27. J.P. Bergfield, M.A. Ratner, C.A. Stafford, M. Di Ventra, Phys. Rev. B **91**, 125407 (2015). https://doi.org/10.1103/PhysRevB.91.125407. http://link.aps.org/doi/10.1103/PhysRevB.91.125407
28. L. Ye, D. Hou, X. Zheng, Y. Yan, M. Di Ventra, Phys. Rev. B **91**, 205106 (2015). https://doi.org/10.1103/PhysRevB.91.205106. http://link.aps.org/doi/10.1103/PhysRevB.91.205106
29. Y. Chen, M. Zwolak, M. Di Ventra, Nano Lett. **3**, 1691 (2003)
30. Y. Ming, Z.X. Wang, Z.J. Ding, H.M. Li, New J. Phys. **12**, 103041 (2010)
31. M. Galperin, A. Nitzan, M.A. Ratner, Phys. Rev. B **75**, 155312 (2007). https://doi.org/10.1103/PhysRevB.75.155312. http://link.aps.org/doi/10.1103/PhysRevB.75.155312
32. Y. de Wilde, F. Formanek, R. Carminati, B. Gralak, P.A. Lemoine, K. Joulain, J.P. Mulet, Y. Chen, J.J. Greffet, Nature **444**, 740 (2006). https://doi.org/10.1038/nature05265
33. Y. Yue, J. Zhang, X. Wang, Small **7**(23), 3324 (2011)
34. J.J. Greffet, C. Henkel, Contemp. Phys. **48**(4), 183 (2007). https://doi.org/10.1080/00107510701690380
35. J. Meair, J.P. Bergfield, C.A. Stafford, P. Jacquod, Phys. Rev. B **90**, 035407 (2014). https://doi.org/10.1103/PhysRevB.90.035407. http://link.aps.org/doi/10.1103/PhysRevB.90.035407
36. Y. Meir, N.S. Wingreen, Phys. Rev. Lett. **68**, 2512 (1992)
37. J.P. Bergfield, C.A. Stafford, Nano Lett. **9**, 3072 (2009)
38. G. Stefanucci, R. van Leeuwen, *Nonequilibrium Many-Body Theory of Quantum Systems: A Modern Introduction* (Cambridge University Press, Cambridge, 2013)
39. R.A. Horn, C.R. Johnson (eds.), *Matrix Analysis* (Cambridge University Press, New York, 1986)
40. J.P. Bergfield, C.A. Stafford, Phys. Rev. B **90**, 235438 (2014). https://doi.org/10.1103/PhysRevB.90.235438. http://link.aps.org/doi/10.1103/PhysRevB.90.235438
41. C.J. Chen, *Introduction to Scanning Tunneling Microscopy*, 2nd edn. (Oxford University Press, New York, 1993)
42. S.V. Kalinin, A. Gruverman, *Scanning Probe Microscopy: Electrical and Electromechanical Phenomena at the Nanoscale* (Springer, Berlin, 2007). https://doi.org/10.1007/978-0-387-28668-6
43. H. Pothier, S. Guéron, N.O. Birge, D. Esteve, M.H. Devoret, Phys. Rev. Lett. **79**, 3490 (1997). https://doi.org/10.1103/PhysRevLett.79.3490. https://link.aps.org/doi/10.1103/PhysRevLett.79.3490
44. L. Onsager, Phys. Rev. **37**, 405 (1931). https://doi.org/10.1103/PhysRev.37.405. http://link.aps.org/doi/10.1103/PhysRev.37.405
45. U. Sivan, Y. Imry, Phys. Rev. B **33**, 551 (1986). https://doi.org/10.1103/PhysRevB.33.551. http://link.aps.org/doi/10.1103/PhysRevB.33.551
46. M. Büttiker, IBM J. Res. Dev. **32**, 63 (1988)

47. M. Büttiker, Phys. Rev. B **40**(5), 3409 (1989). https://doi.org/10.1103/PhysRevB.40.3409
48. A.D. Benoit, S. Washburn, C.P. Umbach, R.B. Laibowitz, R.A. Webb, Phys. Rev. Lett. **57**, 1765 (1986). https://doi.org/10.1103/PhysRevLett.57.1765. http://link.aps.org/doi/10.1103/PhysRevLett.57.1765
49. K.L. Shepard, M.L. Roukes, B.P. van der Gaag, Phys. Rev. B **46**, 9648 (1992). https://doi.org/10.1103/PhysRevB.46.9648. http://link.aps.org/doi/10.1103/PhysRevB.46.9648
50. R. de Picciotto, H.L. Stormer, L.N. Pfeiffer, K.W. Baldwin, K.W. West, Nature **411**(6833), 51 (2001). http://dx.doi.org/10.1038/35075009
51. B. Gao, Y.F. Chen, M.S. Fuhrer, D.C. Glattli, A. Bachtold, Phys. Rev. Lett. **95**, 196802 (2005). https://doi.org/10.1103/PhysRevLett.95.196802. http://link.aps.org/doi/10.1103/PhysRevLett.95.196802
52. M. Büttiker, Phys. Rev. B **32**, 1846 (1985). https://doi.org/10.1103/PhysRevB.32.1846. http://link.aps.org/doi/10.1103/PhysRevB.32.1846
53. M. Büttiker, Phys. Rev. B **33**, 3020 (1986). https://doi.org/10.1103/PhysRevB.33.3020. http://link.aps.org/doi/10.1103/PhysRevB.33.3020
54. J.L. D'Amato, H.M. Pastawski, Phys. Rev. B **41**, 7411 (1990). https://doi.org/10.1103/PhysRevB.41.7411. http://link.aps.org/doi/10.1103/PhysRevB.41.7411
55. T. Ando, Surf. Sci. **361362**, 270 (1996). http://dx.doi.org/10.1016/0039-6028(96)00400-1. http://www.sciencedirect.com/science/article/pii/0039602896004001; Proceedings of the Eleventh International Conference on the Electronic Properties of Two-Dimensional Systems
56. D. Roy, A. Dhar, Phys. Rev. B **75**, 195110 (2007). https://doi.org/10.1103/PhysRevB.75.195110. http://link.aps.org/doi/10.1103/PhysRevB.75.195110
57. M. de Jong, C. Beenakker, Physica A **230**(1–2), 219 (1996). http://www.sciencedirect.com/science/article/pii/0378437196000684
58. S.A. van Langen, M. Büttiker, Phys. Rev. B **56**, R1680 (1997). https://doi.org/10.1103/PhysRevB.56.R1680. http://link.aps.org/doi/10.1103/PhysRevB.56.R1680
59. H. Förster, P. Samuelsson, S. Pilgram, M. Büttiker, Phys. Rev. B **75**, 035340 (2007). https://doi.org/10.1103/PhysRevB.75.035340. http://link.aps.org/doi/10.1103/PhysRevB.75.035340
60. K. Saito, G. Benenti, G. Casati, T.c.v. Prosen, Phys. Rev. B **84**, 201306 (2011). https://doi.org/10.1103/PhysRevB.84.201306. http://link.aps.org/doi/10.1103/PhysRevB.84.201306
61. V. Balachandran, G. Benenti, G. Casati, Phys. Rev. B **87**, 165419 (2013). https://doi.org/10.1103/PhysRevB.87.165419. http://link.aps.org/doi/10.1103/PhysRevB.87.165419
62. K. Brandner, K. Saito, U. Seifert, Phys. Rev. Lett. **110**, 070603 (2013). https://doi.org/10.1103/PhysRevLett.110.070603. http://link.aps.org/doi/10.1103/PhysRevLett.110.070603
63. M. Bandyopadhyay, D. Segal, Phys. Rev. E **84**, 011151 (2011). https://doi.org/10.1103/PhysRevE.84.011151. http://link.aps.org/doi/10.1103/PhysRevE.84.011151
64. S. Bedkihal, M. Bandyopadhyay, D. Segal, The Eur. Phys. J. B **86**(12), 506 (2013). https://doi.org/10.1140/epjb/e2013-40971-7
65. S. Bedkihal, M. Bandyopadhyay, D. Segal, Phys. Rev. B **88**, 155407 (2013). https://doi.org/10.1103/PhysRevB.88.155407. http://link.aps.org/doi/10.1103/PhysRevB.88.155407
66. R. Clausius, Ann. Phys. **169**(12), 481 (1854). https://doi.org/10.1002/andp.18541691202. http://dx.doi.org/10.1002/andp.18541691202
67. J.P. Bergfield, M.A. Solis, C.A. Stafford, ACS Nano **4**(9), 5314 (2010)
68. L. Onsager, Phys. Rev. **38**, 2265 (1931). https://doi.org/10.1103/PhysRev.38.2265. https://link.aps.org/doi/10.1103/PhysRev.38.2265
69. R. Pathria, P. Beale, *Statistical Mechanics, 3rd edn.* (Elsevier, Butterworth-Heinemann, Oxford, 2011)

Chapter 3
Coldest Measurable Temperature

3.1 Introduction

In Chap. 2 we found that the second law of thermodynamics imposes strong restrictions on what can be considered a meaningful thermodynamic measurement. We ask a question motivated by the third law of thermodynamics: What is the coldest possible temperature one can measure in a nonequilibrium quantum system? We have discussed how to measure temperature and voltage in the previous chapter. Most importantly, we realized that temperature and voltage have to be measured simultaneously to ensure uniqueness of the measurement.

Consider the heat current flowing into the probe reservoir p as written in Eq. (2.2)

$$I_p^{(1)} = \frac{1}{h} \int_{-\infty}^{\infty} d\omega (\omega - \mu_p) \mathcal{T}_{ps}(\omega)[f_s(\omega) - f_p(\omega)]. \qquad (3.1)$$

It can be seen that as $T_p \to 0$, the heat current is always positive since $\mathcal{T}_{ps}(\omega) \geq 0$ and $0 \leq f_s \leq 1$

$$\lim_{T_p \to 0} I_p^{(1)} = \frac{1}{h} \int_{-\infty}^{\mu_p} d\omega (\omega - \mu_p) \mathcal{T}_{ps}(\omega)[f_s(\omega) - 1]$$

$$+ \frac{1}{h} \int_{\mu_p}^{\infty} d\omega (\omega - \mu_p) \mathcal{T}_{ps}(\omega) f_s(\omega) \qquad (3.2)$$

$$> 0,$$

since both terms on the r.h.s. are positive and neither can be zero due to our original Postulate 2.1 about the measurability of $\mathcal{T}_{ps}$ and f_s. This is a statement of the third law of thermodynamics which tells us that it is impossible to cool a system to zero temperature. In fact, it tells us that we will necessarily heat a system at zero

© Springer Nature Switzerland AG 2019

A. Shastry, *Theory of Thermodynamic Measurements of Quantum Systems Far from Equilibrium*, Springer Theses, https://doi.org/10.1007/978-3-030-33574-8_3

temperature. It is also evident from Eq. (3.2) that one cannot obtain a measurement result of $T_p = 0$ by solving $I_p^{(\nu)} = 0$ for vanishing particle ($\nu = 0$) and heat currents ($\nu = 1$).

Suppose all the reservoirs are at zero temperature and there is no voltage bias. This of course corresponds to an equilibrium system at zero temperature and there is no transport. In this case, the heat current in Eq. (3.2) goes to zero and we obtain the measurement result to be $T_p = 0$. Even the slightest voltage bias will in fact render the measurement $T_p > 0$ and thus it is impossible to measure $T_p = 0$ for nonequilibrium systems. However, we can still ask the question as to what the coldest possible temperature is in a system out of equilibrium. Clearly, applying a voltage bias at zero temperature leads to Joule heating and there can be no thermoelectric cooling to counter it. Therefore we select an idealized situation where one reservoir is at or near absolute zero and another is at finite temperature and there is no voltage bias. Can there then be a local point in the system where a probe can measure zero temperature? We present a detailed analysis [1] of this question in the present chapter.

3.1.1   Local Temperature and Voltage Measurements

We briefly review in this section the definitions of a temperature and voltage measurement. This was the primary concern of Chap. 2 and the reader may skip this section entirely if they are aware of the results from Chap. 2 or may refer to Chap. 2 if further clarifications are needed. This section has been added merely to let the chapter stand on its own. The main concern of this chapter is a detailed analysis of the third law scenario. A brief reminder of local temperature and voltage measurements is given below but the reader may also refer to Sect. 2.3 and its preceding discussion in Sect. 2.2.

The Green's function formalism provides the following general formula for the particle and heat currents flowing into a reservoir p:

$$I_p^{(\nu)} = -\frac{i}{h} \int_{-\infty}^{\infty} d\omega (\omega - \mu_p)^\nu \, \text{Tr} \left\{ \Gamma^p(\omega) \left(G^<(\omega) + f_p(\omega) \left[G^>(\omega) - G^<(\omega) \right] \right) \right\},$$

(3.3)

which is valid for arbitrary interactions and for transport arbitrarily far from equilibrium.

A definition for a local electron temperature and voltage measurement on the system that takes into account the thermoelectric corrections was proposed in Ref. [2] by noting that the temperature T_p and chemical potential μ_p should be simultaneously defined by the requirement that both the electric current and the electronic heat current into the probe vanish:

$$I_p^{(\nu)} = 0, \quad \nu \in \{0, 1\}.$$

(3.4)

Equation (3.4) gives the conditions under which the probe is in local equilibrium with the sample, which is itself arbitrarily far from equilibrium.

Previous analyses [2–5] have considered this problem within linear response theory, which reduces the system of nonlinear equations (3.4) to a system of equations linear in T_p and μ_p. In this chapter, we consider a problem that is essentially outside the linear response regime and solve the nonlinear system of equations (3.4) numerically. One reservoir at zero temperature and another at a finite temperature entirely invalidate the expressions obtained within the linear response regime which in turn rely upon the Taylor series expansions of the Fermi function.

It was shown in Ref. [6] that Eq. (3.3) can be written in terms of the local properties of the nonequilibrium system. The mean local spectrum sampled by the probe was defined as

$$\bar{A}(\omega; \mathbf{x}) \equiv \frac{\text{Tr}\{\Gamma^P(\omega; \mathbf{x})A(\omega)\}}{\text{Tr}\{\Gamma^P(\omega; \mathbf{x})\}}, \qquad (3.5)$$

where $\mathbf{x}$ is the position of the probe and $A(\omega) = i\big(G^r(\omega) - G^a(\omega)\big)/2\pi$ is the spectral function of the nonequilibrium system. Motivated by the relation at equilibrium, $G_{eq}^<(\omega) = 2\pi i A(\omega) f_{eq}(\omega)$, the local nonequilibrium distribution function (sampled by the probe) was defined as

$$f_s(\omega; \mathbf{x}) \equiv \frac{\text{Tr}\{\Gamma^P(\omega; \mathbf{x})G^<(\omega)\}}{2\pi i \, \text{Tr}\{\Gamma^P(\omega; \mathbf{x})A(\omega)\}}. \qquad (3.6)$$

The mean local occupancy of the system orbitals sampled by the probe is [6]

$$\langle N(\mathbf{x})\rangle = \int_{-\infty}^{\infty} d\omega \bar{A}(\omega) f_s(\omega), \qquad (3.7)$$

and similarly, the mean local energy of the system orbitals sampled by the probe is [6]

$$\langle E(\mathbf{x})\rangle = \int_{-\infty}^{\infty} d\omega \, \omega \bar{A}(\omega) f_s(\omega). \qquad (3.8)$$

Equations (3.5) and (3.6) allow us to rewrite Eq. (3.3) in a form analogous to the two-terminal Landauer–Büttiker formula

$$I_p^{(\nu)} = \frac{1}{h} \int_{-\infty}^{\infty} d\omega (\omega - \mu_p)^\nu 2\pi \, \text{Tr}\{\Gamma^P(\omega)A(\omega)\}[f_s(\omega) - f_p(\omega)]. \qquad (3.9)$$

It was noted that, for the case of maximum local coupling, $[\Gamma^P(\omega)]_{ij} = \Gamma^P(\omega)\delta_{in}\delta_{jn}$, where i, j, and n label states in the one-particle Hilbert space of the nanostructure, the quantities $\bar{A}(\omega) = A_{nn}(\omega)$ and $f_s(\omega)$ become independent of the probe coupling, and can be related by the familiar equilibrium-type formula, $G_{nn}^{<}(\omega) = 2\pi i A_{nn}(\omega) f_s(\omega)$, even though the system is out of equilibrium.

In the broadband limit $\Gamma^p(\omega) \approx \Gamma^p(\mu_0)$, where μ_0 is the equilibrium Fermi energy of the system, we may write $\mathrm{Tr}\{\Gamma^p(\mu_0)\} = \bar{\Gamma}^p$ so that $\bar{A}(\omega) = \mathrm{Tr}\{\Gamma^p(\mu_0) A(\omega)\}/\bar{\Gamma}^p$. From Eq. (3.9), we have

$$I_p^{(\nu)} = \frac{\bar{\Gamma}^p}{\hbar} \int_{-\infty}^{\infty} d\omega (\omega - \mu_p)^\nu \bar{A}(\omega) [f_s(\omega) - f_p(\omega)]. \tag{3.10}$$

It was noted that the equilibrium condition of Eq. (3.4) now implies that the mean local occupancy and energy of the nonequilibrium system are the same as if its nonequilibrium spectrum $\bar{A}(\omega)$ were populated by the equilibrium Fermi–Dirac distribution of the probe $f_p(\omega)$:

$$\langle N \rangle \big|_{f_p} = \langle N \rangle \big|_{f_s} \tag{3.11}$$

$$\langle E \rangle \big|_{f_p} = \langle E \rangle \big|_{f_s}, \tag{3.12}$$

i.e., the probe equilibrates with the system in such a way that $f_p(\omega)$ satisfies the constraints imposed by Eqs. (3.11) and (3.12). A local temperature (and voltage) measurement therefore reveals information about the local energy and charge distribution of the system. The existence and uniqueness of the distribution $f_p(\omega)$ were shown [7] in Chap. 2 and important connections to the second law of thermodynamics were established. We also developed the notion of an ideal probe in Chap. 2 and noted that such a probe operates in the broadband limit (see discussion in Sect. 2.5.2).

3.2   Local Temperatures Near Absolute Zero

Using arguments involving NEGF identities (see e.g., Ref. [8]), one can show that the total current into the probe is the sum of an elastic contribution and an inelastic contribution:

$$I_p^{(\nu)} = I_p^{(\nu)} \Big|_{el} + I_p^{(\nu)} \Big|_{in}. \tag{3.13}$$

The elastic contribution to the current is [9, 10]

$$I_p^{(\nu)} \Big|_{el} = \frac{1}{h} \sum_\alpha \int_{-\infty}^{\infty} d\omega \, (\omega - \mu_p)^\nu \, \mathcal{T}_{p\alpha}(\omega) \left[f_\alpha(\omega) - f_p(\omega) \right], \tag{3.14}$$

where the elastic transmission function is given by [11–14]

$$\mathcal{T}_{p\alpha}(\omega) = \text{Tr}\left\{ \Gamma^p(\omega) G^r(\omega) \Gamma^\alpha(\omega) G^a(\omega) \right\}. \tag{3.15}$$

The inelastic contribution to the current is

$$
\begin{aligned}
I_p^{(\nu)}\Big|_{\text{in}} = &-\frac{i}{h} \int_{-\infty}^{\infty} d\omega (\omega - \mu_p)^\nu \\
&\times \text{Tr}\left\{ \Gamma^p G^r \left[(1 - f_p) \Sigma_{\text{in}}^< + f_p \Sigma_{\text{in}}^> \right] G^a \right\},
\end{aligned} \tag{3.16}
$$

where Σ_{in} is the self-energy due to electron–electron and electron–phonon interactions.

Our analyses in this chapter consider a quantum conductor that is placed in contact with two electron reservoirs: a cold reservoir $R1$ and a hot reservoir $R2$. We are interested in the limiting case where reservoir $R1$ is held near absolute zero ($T_1 \to 0$) while $R2$ is held at finite temperature ($T_2 = 100\,\text{K}$ in our simulations). We assume no electrical bias ($\mu_1 = \mu_2 = \mu_0$).

Transport in this regime occurs within a narrow thermal window close to the Fermi energy of the reservoirs, and will thus be dominated by elastic processes, so that the contribution to the probe current from Eq. (3.16) can be neglected. Typically,

$$I_p^{(\nu)}\Big|_{\text{in}} \Big/ I_p^{(\nu)}\Big|_{\text{el}} \sim \mathcal{O}(k_B T / \Delta E)^2 \tag{3.17}$$

due to the suppression of inelastic electron scattering by the Pauli exclusion principle at low temperatures, where ΔE is a characteristic electronic energy scale of the problem: $\Delta E = \varepsilon_F$ for bulk systems, $\Delta E = \max\{\Gamma, \varepsilon_F - \varepsilon_{\text{res}}\}$ in the resonant tunneling regime. It should be noted that, although the transport energy window is small, this problem is essentially outside the scope of linear response theory owing to the large discrepancies in the derivatives of the Fermi functions of the two reservoirs. Therefore, the problem has to be addressed with the full numerical evaluation of the currents given by Eq. (3.14).

A pure thermal bias, such as the one considered in this chapter, has been shown to lead to temperature oscillations in small molecular junctions [2] and 1-D conductors [15, 16]. Temperature oscillations have been predicted in quantum coherent conductors such as graphene [5], which allow the oscillations to be tuned (e.g., by suitable gating) such that they can be resolved under existing experimental techniques [17–20] of SThM. More generally, quantum coherent temperature oscillations can be obtained for quantum systems driven out of equilibrium due to external fields [21, 22] as well as chemical potential [4] and temperature bias of the reservoirs. In practice, the thermal coupling of the probe to the environment sets limitations on the resolution of a scanning thermoelectric probe [2]. However, in this chapter, we ignore the coupling of the probe to the ambient environment, in order to highlight the theoretical limitations on temperature measurements near absolute zero.

In evaluating the expressions for the currents in Eq. (3.14) within elastic transport theory, we encounter integrals of the form $\int_{-\infty}^{\infty} d\omega F(\omega)\left(f_2(\omega) - f_1(\omega)\right)$. We use the Sommerfeld series given by

$$\int_{-\infty}^{\infty} d\omega F(\omega)\big(f_2(\omega) - f_1(\omega)\big) = \int_{\mu_1}^{\mu_2} d\omega F(\omega)$$

$$+ 2\sum_k \Theta(k+1)\big[(k_B T_2)^{k+1} F^{(k)}(\mu_2) - (k_B T_1)^{k+1} F^{(k)}(\mu_1)\big], \qquad (3.18)$$

$$k \in \{1, 3, 5, \ldots\},$$

where we use the symbol Θ that relates to the Riemann zeta function ζ as

$$\Theta(k+1) = \left(1 - \frac{1}{2^k}\right)\zeta\left(k+1\right) \qquad (3.19)$$

and $f_\alpha(\omega)$ is the Fermi–Dirac distribution of reservoir α. The second term on the r.h.s. of Eq. (3.18) accounts for the exponential tails in $\big(f_2(\omega) - f_1(\omega)\big)$, and its contribution depends on the changes to the function $F(\omega)$ in the neighborhoods of $\omega = \mu_1$ and $\omega = \mu_2$, and can generally be truncated using a Taylor series expansion for most well-behaved functions $F(\omega)$. The l.h.s. of Eq. (3.18) is bounded if $F(\omega)$ grows slower than exponentially for $\omega \to \pm\infty$ and is satisfied by the current integrals in Eq. (3.14).

3.2.1   Constant Transmissions

In order to make progress analytically, we consider first the case of constant transmissions:

$$\mathcal{T}_{p\alpha}(\omega) = \mathcal{T}_{p\alpha}(\mu_0) \equiv \mathcal{T}_{p\alpha}. \qquad (3.20)$$

This is a reasonable assumption because the energy window involved in transport is of the order of the thermal energy of the hot reservoir ($k_B T_2 \approx 25\,\mathrm{meV}$, at room temperature) and we may expect no great changes to the transmission function for a small system. In this case the series (3.18) for the number current contains no temperature terms at all, while the heat current contains terms quadratic in the temperature. It is easy to see that the expression for the number current into the probe becomes

$$I_p^{(0)} = \frac{1}{h}\sum_\alpha \mathcal{T}_{p\alpha}(\mu_\alpha - \mu_p), \qquad (3.21)$$

and the heat current into the probe is given by

$$I_p^{(1)} = \frac{1}{h} \sum_\alpha \left(\mathcal{T}_{p\alpha} \frac{(\mu_\alpha - \mu_p)^2}{2} + \frac{\pi^2 k_B^2}{6} \mathcal{T}_{p\alpha} (T_\alpha^2 - T_p^2) \right). \tag{3.22}$$

Equation (3.21) does not depend on the temperature and can be solved readily:

$$\mu_p = \mu_0, \tag{3.23}$$

since $\mu_1 = \mu_2 = \mu_0$ and Eq. (3.22) is solved by

$$T_p = \sqrt{\frac{\mathcal{T}_{p1} T_1^2 + \mathcal{T}_{p2} T_2^2}{\mathcal{T}_{p1} + \mathcal{T}_{p2}}}. \tag{3.24}$$

In this chapter, we are primarily interested in temperature measurements near absolute zero and work in the limit $T_1 \to 0$ which yields

$$T_p = \sqrt{\frac{\mathcal{T}_{p2}}{\mathcal{T}_{p1} + \mathcal{T}_{p2}}} T_2. \tag{3.25}$$

We have $T_p \to 0$ as $\mathcal{T}_{p2} \to 0$. Indeed, when the system is decoupled from the hot reservoir $R2$, the probe would read the temperature of reservoir $R1$.

3.2.2 Transmission Node

The analysis of the previous section suggests that a suppression in the transmission from the finite-temperature reservoir $R2$ results in probe temperatures in the vicinity of absolute zero. In quantum coherent conductors, destructive interference gives rise to nodes in the transmission function. In this section, we consider a case where the transmission from $R2$ into the probe has a node at the Fermi energy. In the vicinity of such a node, generically, the transmission probability varies quadratically with energy:

$$\mathcal{T}_{p2}(\omega) = \frac{1}{2} \mathcal{T}_{p2}^{(2)} (\omega - \mu_0)^2, \tag{3.26}$$

while the transmission from the cold reservoir $R1$ may still be treated as a constant:

$$\mathcal{T}_{p1}(\omega) = \mathcal{T}_{p1}. \tag{3.27}$$

Applying the Sommerfeld series (3.18) for the number current gives us

$$I_p^{(0)} = \frac{1}{h}\left(\mathcal{T}_{p1}(\mu_0 - \mu_p) - \frac{\mathcal{T}_{p2}^{(2)}}{6}(\mu_p - \mu_0)^3 \right.$$
$$\left. + \frac{\pi^2}{6}\mathcal{T}_{p2}^{(2)}(\mu_0 - \mu_p)k_B^2 T_p^2 \right), \tag{3.28}$$

where the $k_B T_2$ term is still missing since the first derivative of $\mathcal{T}_{p2}(\omega)$ vanishes at $\mu_2 = \mu_0$. We note that Eq. (3.28) admits a single real root at

$$\mu_p = \mu_0. \tag{3.29}$$

With this solution, we can write down the equation for the heat current as

$$I_p^{(1)} = \frac{1}{h}\left(\frac{\pi^2 k_B^2}{6}\mathcal{T}_{p1}(T_1^2 - T_p^2) + \frac{7}{8}\frac{\pi^4 k_B^4}{15}\mathcal{T}_{p2}^{(2)}(T_2^4 - T_p^4) \right), \tag{3.30}$$

which gives us a simple quadratic equation in T_p^2. We note that the above equation is monotonically decreasing in T_p for all positive values of temperature. There exists a unique solution to Eq. (3.30) in the interval $T_1 < T_p < T_2$, since $I_p^{(1)}(T_p)$ undergoes a sign change between these two values, and is also the only positive solution due to monotonicity. Physically, this solution is reasonable since we expect a temperature measurement to be within the interval (T_1, T_2) in the absence of an electrical bias. It is straightforward to write down the exact solution[1] to Eq. (3.30). However, we simplify the expression for T_p by noting that

$$\mathcal{T}_{p2}^{(2)}(k_B T_2)^2 \ll \mathcal{T}_{p1}, \tag{3.31}$$

that is, the transmission into the probe from $R2$ within a thermal energy window $k_B T_2$ in the presence of a node is small in comparison to the transmission from $R1$. The approximate solution for T_p then becomes

---

[1] The exact solution to the Eq. (3.30) with $T_1 \rightarrow 0$ is

$$T_p = T_2\left(\frac{\sqrt{1 + 4\lambda_1^2} - 1}{2\lambda_1} \right)^{\frac{1}{2}},$$

where λ_1 is defined in Eq. (3.39) and simplifies to

$$\lambda_1 = \frac{7\pi^2}{20}\frac{\mathcal{T}_{p2}^{(2)}(k_B T_2)^2}{\mathcal{T}_{p1}},$$

a factor that appears in Eq. (3.32).

$$T_p = \sqrt{\frac{7\pi^2}{20} \frac{T_{p2}^{(2)}(k_B T_2)^2}{T_{p1}}} T_2, \tag{3.32}$$

where, as before, we have taken the limiting case where $T_1 \to 0$.

3.2.3 Higher-Order Destructive Interference

Although a generic node obtained in quantum coherent transport depends quadratically on the energy, it is possible to obtain higher-order "supernodes" in some systems [14]. In the vicinity of such a supernode, the transmission function can be written as

$$T_{p2}(\omega) = \frac{1}{2n!} T_{p2}^{(2n)}\big|_{\omega=\mu_0}(\omega - \mu_0)^{2n}, \tag{3.33}$$

while the transmission from $R1$ may still be approximated by Eq. (3.27). Exact expressions for the currents can again be evaluated using Eq. (3.18). The expression for the number current becomes

$$
\begin{aligned}
I_p^{(0)} = \frac{1}{h}\Bigg(& T_{p1}(\mu_0 - \mu_p) - \frac{T_{p2}^{(2n)}}{(2n+1)!}(\mu_p - \mu_0)^{2n+1} \\
& + 2\sum_{k\in odd}\Theta(k+1)\Big[(k_B T_2)^{k+1}T_{p2}^{(k)}(\mu_0) - (k_B T_p)^{k+1}T_{p2}^{(k)}(\mu_p)\Big]\Bigg),
\end{aligned}
\tag{3.34}
$$

and we have

$$T_{p2}^{(k)}(\mu_0) = 0, \quad \forall k \in \{1, 3, 5, \ldots\}. \tag{3.35}$$

Now, with

$$\mu_p = \mu_0, \tag{3.36}$$

every term on the r.h.s. of Eq. (3.34) vanishes. Taking $\mu_p = \mu_0$, we proceed to write the equation for the heat current. Using Eq. (3.18) with $F(\omega) = (\omega - \mu_0)T_{p\alpha}(\omega)$, we obtain only one nonvanishing derivative for each reservoir, that is, $F^{(1)}(\mu_0) = T_{p1}(\mu_0)$ for $R1$ and $F^{(2n+1)}(\mu_0) = (2n+1)T_{p2}^{(2n)}$ for $R2$. Therefore,

$$
\begin{aligned}
I_p^{(1)} = \frac{2}{h}\Bigg(& \Theta(2)T_{p1}\Big[(k_B T_1)^2 - (k_B T_p)^2\Big] \\
& + (2n+1)\Theta(2n+2)T_{p2}^{(2n)}\Big[(k_B T_2)^{2n+2} - (k_B T_p)^{2n+2}\Big]\Bigg),
\end{aligned}
\tag{3.37}
$$

which is a polynomial equation in T_p of degree $(2n + 2)$. We can rewrite Eq. (3.37) as a polynomial $p(x)$ in $x = T_p/T_2$:

$$p(x) = x^2 + \lambda_n x^{2n+2} - \lambda_n = 0, \tag{3.38}$$

where we have taken $T_1 \to 0$ for $R1$, and λ_n is a dimensionless quantity given by

$$\lambda_n = (2n + 1)\frac{\Theta(2n + 2)}{\Theta(2)}\left(\frac{T_{p2}^{(2n)}(k_B T_2)^{2n}}{T_{p1}}\right). \tag{3.39}$$

We will have $\lambda_n \ll 1$ for a suitable energy window set by $k_B T_2$, since the transmission into the probe from $R2$ suffers destructive interference at the Fermi energy μ_0. If the thermal energy is large enough, then this approximation may no longer hold. In any case, it is possible to define a temperature T_2 so that this approximation is strongly valid. Under the validity of this approximation, the solution to Eq. (3.38) can be written using perturbation theory as

$$x = \sqrt{\lambda_n}\left(1 + \mathcal{O}(\lambda_n^{n+1})\right), \tag{3.40}$$

with corrections that are of much higher-order in λ_n. The solution for T_p given in Eq. (3.40) reduces to Eq. (3.25) for the case with constant transmissions by setting $n = 0$, and to the approximate result obtained in Eq. (3.32) in the presence of a node by setting $n = 1$. We note that higher-order interference effects cause the probe temperature to decay more rapidly with respect to T_2 since $T_p \sim T_2^{n+1}$, that is, when T_2 is halved, T_p is reduced by a factor of 2^{n+1}. Now, if we consider the limiting case where $R2$ is also cooled to absolute zero, $T_2 \to 0$, Eq. (3.40) implies that $T_p \to 0$ at least as quickly as T_2 (in the absence of destructive interference, i.e., $n = 0$) or quicker (when there is destructive interference, i.e., $n \geq 1$).

3.2.4 Uniqueness and the Second Law

It should be noted that the polynomial given in Eq. (3.38) is monotonically increasing for all positive x and furthermore, there is only one positive root since $p(0) < 0$ and $p(1) > 0$. Stated in terms of T_p, this implies the existence of a unique solution for the measured temperature T_p in the interval $T_1 < T_p < T_2$, as noted in the previous subsection [T_1 has been set to zero in Eq. (3.38)]. The sign of $p(x)$ essentially tells us the direction of heat flow for a temperature bias of the probe with respect to its equilibrium value. $p(x) < 0$ ($p(x) > 0$) corresponds to heat flowing into (out of) the probe, since we changed the sign in re-arranging Eq. (3.37) to Eq. (3.38). The monotonicity of $p(x)$ is therefore equivalent to the Clausius statement of the second law of thermodynamics, and also ensures the uniqueness of the temperature measurement. Such statements of the second law were explored in great detail in Chap. 2.

3.3 Numerical Results and Discussion

In order to illustrate the above theoretical results, and further characterize the local properties of the nonequilibrium steady state, we now present numerical calculations for several molecular junctions with π-conjugation. In all of the simulations, the molecule is connected to a cold reservoir $R1$ at $T_1 = 0$ K and a hot reservoir $R2$ at $T_2 = 100$ K. There is no electrical bias; both electrodes have chemical potential μ_0. The temperature probe is modeled as an atomically sharp Au tip scanned horizontally at a constant vertical height of 3.5Å above the plane of the carbon nuclei in the molecule (tunneling regime). Figures 3.1, 3.2, and 3.3 illustrate our findings for 3 different molecular junctions where we consider several different geometries for the reservoir-system couplings.

The molecular Hamiltonian is described within Hückel theory, $H_{mol} = \sum_{<i,j>} t_{ij} d_i^{\dagger} d_j + $ h.c., with nearest-neighbor hopping matrix element $t = -2.7$ eV. The coupling of the molecule with the reservoirs is described by the tunneling-width matrices Γ^{α}. The retarded Green's function of the junction is given by $G^r(\omega) = [\mathbb{S}\omega - H_{mol} - \Sigma_T(\omega)]^{-1}$, where $\Sigma_T = -i \sum_{\alpha} \Gamma^{\alpha}/2$ is the tunneling self-energy. We take the lead-molecule couplings in the broadband limit, i.e., $\Gamma_{nm}^{\alpha}(\omega) = \Gamma_{nm}^{\alpha}(\mu_0)$ where μ_0 is the Fermi energy of the metal leads. We also take the lead-molecule couplings to be diagonal matrices $\Gamma_{nm}^{\alpha}(\omega) = \Gamma_{\alpha}\delta_{nl}\delta_{ml}$ coupled to a single π-orbital l of the molecule. $\mathbb{S}$ is the overlap-matrix between the atomic orbitals on different sites and we take $\mathbb{S} = \mathbb{1}$, i.e., an orthonormal set of atomic orbitals. The lead-molecule couplings are taken to be symmetric, with $\Gamma_1 = \Gamma_2 = 0.5$ eV. The non-zero elements of the system-reservoir couplings for $R1$ (cold) and $R2$ (hot) are indicated with a blue and red square, respectively, corresponding to the carbon atoms in the molecule covalently bonded to the reservoirs. The tunneling-width matrix Γ^p describing probe-sample coupling is also treated in the broadband limit (see Sect. 3.3.2 for details pertaining to the modeling of probe-system coupling). The probe is in the tunneling regime and the probe-system coupling is weak (few meV) in comparison to the system-reservoir couplings ($\Gamma_1 = \Gamma_2 = 0.5$ eV).

It must be emphasized that, although we take a noninteracting Hamiltonian for the isolated molecule, our results depend only upon the existence of transmission nodes in the elastic transport regime, which are a characteristic feature of coherent quantum transport, and do *not* depend on the particular form of the junction Hamiltonian.

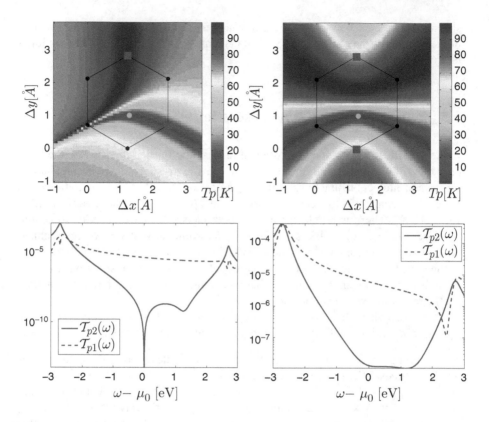

Fig. 3.1 Upper panels: Local temperature distributions for Au-benzene-Au junctions in meta and para configurations, respectively. The thermal bias is supplied by cold ($T_1 = 0\,\mathrm{K}$) and hot ($T_2 = 100\,\mathrm{K}$) reservoirs covalently bonded to the atoms indicated by the blue and red squares, respectively. There is no electrical bias. The probe is scanned at a height of 3.5 Å above the plane of the carbon nuclei in the molecule. The green dots shown in the temperature distributions correspond to the coldest temperature found in each of the junction configurations. Bottom panels: Transmission probabilities into the probe from $R1$ (cold, i.e., blue curve) and $R2$ (hot, i.e., red curve), when the probe is positioned over the coldest spot (shown by the corresponding green dot in the upper panel). A transmission node in the meta configuration leads to a greatly suppressed probe temperature [cf. Table 3.1]. Note the very different vertical scales in the bottom panels

3.3.1 Local Temperatures

We considered several different molecules and electrode configurations, with and without transmission nodes, and searched for the coldest spot in each system (indicated by a green dot in the figures) as measured by the scanning thermoelectric probe.

The local temperature $T_p(\mathbf{x})$ depends on the transmissions from the reservoirs into the probe Eq. (C.13), determined by the local probe-system coupling $\Gamma^p(\mathbf{x})$ (see Sect. 3.3.2), and is thus a function of probe position. The coldest spot was

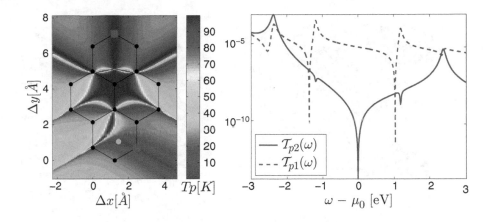

Fig. 3.2 Left panel: Probe temperature distribution in a Au-pyrene-Au junction under the same conditions described in Fig. 3.1. The green dot corresponds to the coldest temperature found by the search algorithm. Right panel: Transmissions into the probe from the hot reservoir $R2$ (red) and the cold reservoir $R1$ (blue) at the coldest position, indicated by the green dot on the left. The probe transmission from $R2$ exhibits a (mid-gap) node at the Fermi energy μ_0 of the reservoirs, thereby suppressing the temperature measured by the probe

found using a particle swarm optimization technique that minimizes the ratio of the transmissions to the probe (within a thermal window) from the hot reservoir $R2$ to that of the cold reservoir $R1$, within a search space that spans the z-plane at 3.5Å and restricted in the xy direction to within 1Å from the edge of the molecule.[2] The numerical solution to Eq. (3.4) was found using Newton's method. While the algorithm was found to converge rapidly for most points (less than 15 iterations), it is still computationally intensive since the evaluation of the currents given by Eq. (3.14) must have sufficient numerical accuracy. We also note that the minimum probe temperature obtained for each junction does not depend strongly on the distance between the plane of the scanning probe and that of the molecule. This is explained as follows: the probe temperature must depend upon the relative magnitudes of the transmissions into the probe from the two reservoirs and not their actual values. Therefore, the temperature remains roughly independent of the coupling strength $\mathrm{Tr}\{\Gamma^p\}$. It has also been previously noted in Ref. [3] that the local temperature measurement showed little change with the coupling strength even when varied over several orders of magnitude. The restrictions placed on our search space within the optimization algorithm are therefore well justified.

Figure 3.1 shows the temperature distribution for two configurations of Au-benzene-Au junctions with the chemical potentials of the metal leads at the middle of the HOMO-LUMO gap. The mid-gap region is advantageous since (1) the

---

[2]It is necessary to minimize the ratio of transmissions to find the temperature minima since it is computationally prohibitive to calculate the temperatures at various points in the search space, within each iteration of the optimization algorithm.

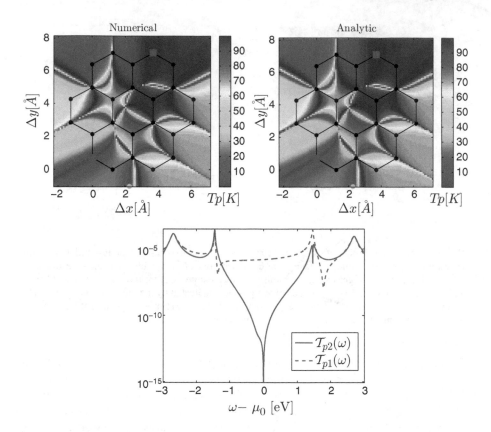

Fig. 3.3 Upper panels: Probe temperature distributions for a Au-coronene-Au junction under the same conditions described in Fig. 3.1. The numerically calculated temperature is on the left, and to the right is the analytically calculated temperature using Eq. (3.25). Although it is in excellent qualitative agreement (and quantitative agreement for the most part), Eq. (3.25) poorly estimates the temperature for the coldest spot (shown in green) due to the existence of a transmission node. Equation (3.32) gives the correct estimation in the presence of a node, while Eq. (3.25) incorrectly predicts $T_p = 0$ K. The lower panel shows the probe transmissions from the two reservoirs ($R1$ with $T_1 = 0$ K in blue and $R2$ with $T_2 = 100$ K in red) corresponding to the probe positioned over the coldest spot (shown in green)

molecule is charge neutral when the lead chemical potentials are tuned to the mid-gap energy and (2) the mismatch between the metal leads' Fermi energy and the mid-gap energy is typically small (less than 1–2 eV for most metal–molecule junctions) and available gating techniques [23] would be sufficient to tune across the gap. In both junctions, the region of lowest temperature passes through the two sites in meta orientation relative to the hot electrode because the transmission probability from the hot electrode into the probe is minimum when it is at these locations [2]. The meta junction exhibits much lower minimum temperature measurements since there is a transmission node from the hot reservoir $R2$ at the mid-gap energy but

Table 3.1 The table shows the lowest temperatures found in the different junctions considered

Junction	$T_p[K]$ numerical	$T_p[K]$ analytic	Eq.
Benzene (meta)	0.154	0.1526	(3.32)
Benzene (para)	4.624	4.627	(3.25)
Pyrene	0.0821	0.0817	(3.32)
Coronene	0.0349	0.0355	(3.32)

All junctions have the same bias conditions: $T_1 = 0\,\mathrm{K}$, $T_2 = 100\,\mathrm{K}$, and no electrical bias. The right-most column shows the equation used to compute the temperature analytically. We obtain excellent agreement between the numerical and analytic results. The para configuration of the benzene junction does not display a node in the probe transmission spectrum and therefore the minimum probe temperature is not strongly suppressed

such nodes are absent in the para junction. Table 3.1 shows the coldest temperature found in each of the junctions presented here.

We only present two junction geometries for benzene here to illustrate that the existence of a node in the probe transmission spectrum, at the mid-gap energy, depends upon the junction geometry. The nodes are also absent in the ortho configuration of benzene and the lowest temperature found in this case is similar to that found in the para configuration.

Figure 3.2 shows the temperature distribution in a gated Au-pyrene-Au junction, and the transmissions into the probe from the two reservoirs at the coldest spot. In general, nodes in the transmission spectrum occur only in a few of the possible junction geometries. As in the benzene junctions, the coldest regions in the pyrene junction pass through the sites to which electron transfer from the hot electrode is blocked by the *rules of covalence* [2] describing bonding in π-conjugated systems. We note that the temperature distribution shown in Fig. 3.2 differs significantly from that shown in Ref. [2] for four important reasons: (i) the junction configuration in Fig. 3.2 is asymmetric while that considered in Ref. [2] was symmetric; (ii) the thermal coupling κ_{p0} of the temperature probe to the ambient environment has been set to zero in Fig. 3.2 to allow for resolution of temperatures very close to absolute zero, while the probe in Ref. [2] was taken to have $\kappa_{p0} = 10^{-4}\kappa_{p0}$, where $\kappa_{p0} = (\pi^2/3)(k_B^2 T/h) = 2.84 \times 10^{-10}$ W/K at $T = 300\,\mathrm{K}$ is the thermal conductance quantum [24, 25]; (iii) the transport in Fig. 3.2 is assumed to take place at the mid-gap energy due to appropriate gating of the junction, while Ref. [2] considered a junction without gating; and (iv) Ref. [2] considered temperature measurements only in the linear response regime while the thermal bias applied in Fig. 3.2 is essentially outside the scope of linear response. Points (ii)–(iv) also differentiate the results for benzene junctions shown in Fig. 3.1 from the linear-response results of Ref. [2].

Figure 3.3 shows the temperature distribution in a gated Au-coronene-Au junction exhibiting a node in the probe transmission spectrum. The junction shown was one of three such geometries to exhibit nodes (10 distinct junction geometries were considered). Again, the coldest regions in the junction pass through the sites to which electron transfer from the hot electrode is blocked by the rules of covalence.

The coronene junction in Fig. 3.3 displays the lowest temperature among all the different junctions considered, with a minimum temperature of $T_p = 35\,\text{mK}$. It should be noted that this temperature would be suppressed by a factor of 100, i.e., $T_p = 350\,\mu\text{K}$ if $R2$ were held at $10\,\text{K}$ due to the quadratic scaling of T_p with respect to the temperature T_2 [cf. Eq. (3.32)]. Higher-order nodes would produce even greater suppression.

3.3.2   Model of Probe-Sample Coupling

The scanning thermoelectric probe is modeled as an atomically sharp Au tip operating in the tunneling regime. The probe tunneling-width matrices[3] may be described in general as $\Gamma_{nm}^p(\omega) = 2\pi V_n V_m^* \rho_p(\omega)$, where $\rho_p(\omega)$ is the local density of states of the apex atom in the probe electrode and V_m, V_n are the tunneling matrix elements between the quasi-atomic wavefunctions of the apex atom in the electrode and the mth, nth π-orbitals in the molecule. We consider the Au tip to be dominated by the s-orbital character and neglect all other contributions. The probe-system coupling is also treated within the broadband approximation. The tunneling-width matrix Γ^p describing the probe-system coupling is in general non-diagonal and is calculated using the methods highlighted in Ref. [26].

3.4   Conclusions

We asked a question motivated by the third law of thermodynamics: Is it possible to measure a local temperature of absolute zero in a nonequilibrium quantum system? A simple analysis of this question immediately tells us that the answer is no, consistent with the third law of thermodynamics. We noted then that voltage biases near zero temperature necessarily lead to heating. We therefore investigated local electronic temperature distributions in nanoscale quantum conductors with one of the reservoirs held at finite temperature and the other held at or near absolute zero, a problem essentially outside the scope of linear response theory.

We obtained local temperatures close to absolute zero when electrons originating from the finite-temperature reservoir undergo destructive quantum interference. The local temperature was computed by numerically solving a nonlinear system of Eqs. (3.4) and (3.14) describing equilibration of a scanning thermoelectric probe with the system, and we obtain excellent agreement with analytic results Eqs. (3.25), (3.32), and (3.40) derived using the Sommerfeld expansion. Our conclusion is that a local temperature equal to absolute zero is impossible in a nonequilibrium quantum system but arbitrarily low finite values are, in principle, possible.

---

[3]I would like to thank Justin Bergfield for modeling the tip-sample coupling.

References

1. A. Shastry, C.A. Stafford, Phys. Rev. B **92**, 245417 (2015). https://doi.org/10.1103/PhysRevB. 92.245417. http://link.aps.org/doi/10.1103/PhysRevB.92.245417
2. J.P. Bergfield, S.M. Story, R.C. Stafford, C.A. Stafford, ACS Nano **7**(5), 4429 (2013). https:// doi.org/10.1021/nn401027u
3. J. Meair, J.P. Bergfield, C.A. Stafford, P. Jacquod, Phys. Rev. B **90**, 035407 (2014). https://doi. org/10.1103/PhysRevB.90.035407. http://link.aps.org/doi/10.1103/PhysRevB.90.035407
4. J.P. Bergfield, C.A. Stafford, Phys. Rev. B **90**, 235438 (2014). https://doi.org/10.1103/ PhysRevB.90.235438. http://link.aps.org/doi/10.1103/PhysRevB.90.235438
5. J.P. Bergfield, M.A. Ratner, C.A. Stafford, M. Di Ventra, Phys. Rev. B **91**, 125407 (2015). https://doi.org/10.1103/PhysRevB.91.125407. http://link.aps.org/doi/10.1103/PhysRevB.91. 125407
6. C.A. Stafford, Phys. Rev. B **93**, 245403 (2016). https://doi.org/10.1103/PhysRevB.93.245403. http://link.aps.org/doi/10.1103/PhysRevB.93.245403
7. A. Shastry, C.A. Stafford, Phys. Rev. B **94**, 155433 (2016). https://doi.org/10.1103/PhysRevB. 94.155433. http://link.aps.org/doi/10.1103/PhysRevB.94.155433
8. J.K. Viljas, J.C. Cuevas, F. Pauly, M. Häfner, Phys. Rev. B **72**(24), 241201(R) (2005)
9. M. Büttiker, Phys. Rev. Lett. **57**, 1761 (1986)
10. U. Sivan, Y. Imry, Phys. Rev. B **33**, 551 (1986). https://doi.org/10.1103/PhysRevB.33.551. http://link.aps.org/doi/10.1103/PhysRevB.33.551
11. S. Datta, *Electronic Transport in Mesoscopic Systems* (Cambridge University Press, Cambridge, 1995)
12. J. Heurich, J.C. Cuevas, W. Wenzel, G. Schön, Phys. Rev. Lett. **88**, 256803 (2002)
13. J.P. Bergfield, C.A. Stafford, Phys. Rev. B **79**(24), 245125 (2009)
14. J.P. Bergfield, M.A. Solis, C.A. Stafford, ACS Nano **4**(9), 5314 (2010)
15. Y. Dubi, M. Di Ventra, Phys. Rev. E **79**, 042101 (2009). https://doi.org/10.1103/PhysRevE.79. 042101. http://link.aps.org/doi/10.1103/PhysRevE.79.042101
16. Y. Dubi, M. Di Ventra, Nano Lett. **9**(1), 97 (2009)
17. Y.J. Yu, M.Y. Han, S. Berciaud, A.B. Georgescu, T.F. Heinz, L.E. Brus, K.S. Kim, P. Kim, Appl. Phys. Lett. **99**(18), 183105 (2011). http://dx.doi.org/10.1063/1.3657515. http://scitation. aip.org/content/aip/journal/apl/99/18/10.1063/1.3657515
18. K. Kim, J. Chung, G. Hwang, O. Kwon, J.S. Lee, ACS Nano **5**(11), 8700 (2011). https://doi. org/10.1021/nn2026325. http://dx.doi.org/10.1021/nn2026325. PMID: 21999681
19. K. Kim, W. Jeong, W. Lee, P. Reddy, ACS Nano **6**(5), 4248 (2012). https://doi.org/10.1021/ nn300774n. http://dx.doi.org/10.1021/nn300774n. PMID: 22530657
20. F. Menges, H. Riel, A. Stemmer, B. Gotsmann, Nano Lett. **12**(2), 596 (2012). https://doi.org/ 10.1021/nl203169t. http://dx.doi.org/10.1021/nl203169t. PMID: 22214277
21. A. Caso, L. Arrachea, G.S. Lozano, Phys. Rev. B **81**(4), 041301 (2010). https://doi.org/10. 1103/PhysRevB.81.041301
22. A. Caso, L. Arrachea, G.S. Lozano, Phys. Rev. B **83**, 165419 (2011). https://doi.org/10.1103/ PhysRevB.83.165419. http://link.aps.org/doi/10.1103/PhysRevB.83.165419
23. H. Song, Y. Kim, Y.H. Jang, H. Jeong, M.A. Reed, T. Lee, Nature **462**(7276), 1039 (2009). http://dx.doi.org/10.1038/nature08639
24. L.G.C. Rego, G. Kirczenow, Phys. Rev. Lett. **81**(1), 232 (1998)
25. L.G.C. Rego, G. Kirczenow, Phys. Rev. B **59**(20), 13080 (1999)
26. C.J. Chen, *Introduction to Scanning Tunneling Microscopy*, 2nd edn. (Oxford University Press, New York, 1993)

Chapter 4
STM as a Thermometer

In Chap. 2 we showed that a measurement of temperature has to be accompanied with a measurement of voltage as well. We discuss here the experimental consequences [1] of the results of Chap. 2.

4.1 Background and Proposed Thermometer

Thermal imaging of nanoscale systems is of crucial importance not only due to its potential applications in future technologies, but also because it can greatly enhance our understanding of heat transport at the smallest scales. In recent years, nanoscale thermometry has found application in a wide range of fields including thermometry in a living cell [2], local control of chemical reactions [3], and temperature mapping of operating electronic devices [4]. Various studies utilize radiation-based techniques such as Raman spectroscopy [5], fluorescence in nanodiamonds [2, 6], and near-field optical microscopy [7]. The spatial resolution of these radiation-based techniques are limited due to optical diffraction and, to overcome this drawback, scanning probe techniques [8] have seen a flurry of activity in recent years [9–11]. Despite the remarkable progress made by scanning probe thermometry, the spatial resolution remains in the 10 nm range.

Since temperature and voltage are both fundamental thermodynamic observables, it is instructive to draw the sharp contrast that exists between the measurement of these two quantities at the nanoscale. Scanning probe potentiometry (STP) [12] is a mature technology and can map local voltage variations with sub-angstrom spatial resolution by operating in the tunneling regime. STP has been used to map the local voltage variations in the vicinity of individual scatterers, interfaces, or boundaries [13–17], providing direct observations of the Landauer dipole [18]. STP has been a useful tool in disentangling the different scattering mechanisms [17], and can map local potential variations due to quantum interference effects [15, 16].

© Springer Nature Switzerland AG 2019   61
A. Shastry, *Theory of Thermodynamic Measurements of Quantum Systems Far from Equilibrium*, Springer Theses, https://doi.org/10.1007/978-3-030-33574-8_4

Similarly, temperature oscillations due to quantum interference effects have been theoretically predicted for various nanosystems out of equilibrium [19–22] but have hitherto remained outside the reach of experiment.

Scanning thermal microscopy (SThM) relies on the measurement of a heat-flux-related signal that can be sensed based on a calibrated thermocouple or electrical resistor [8]. A good thermal contact between the tip and the sample is needed for an appreciable heat-flux and generally implies a measurement in the contact regime, thus limiting the spatial resolution. Previous studies have addressed various technical difficulties in the measurement such as the parasitic heat transfer through air and the loss of spatial resolution due to the formation of a liquid meniscus at the tip-sample interface [23]. A dramatic enhancement in the spatial resolution was later reported by operating the probe in an ultrahigh vacuum (UHV) environment [9], allowing quantitative thermometry with ~10 nm spatial resolution. SThM measurements are typically affected by contact-related artifacts, most importantly by an unknown tip-sample thermal resistance which previous studies have attempted to mitigate [24]. Additionally, in SThM measurements on systems out of equilibrium, the tip-sample thermal conductance must be large compared to the probe-cantilever conductance which is not easy to achieve. A further concern is that the resulting operation would be invasive (due to the strong thermal coupling between the probe and sample) has also been pointed out. We encourage the reader to refer to Ref. [24] for a recent review and extensive reference list. Recently a significant advance was reported in Ref. [11] where the authors propose a method which could potentially eliminate contact-related artifacts. Here the spatial resolution for temperature imaging was reported to be ~7 nm but is still orders of magnitude coarser than routinely achieved for imaging other physical properties.

Here we propose a method to overcome the most daunting technical challenges of SThM by simultaneously measuring the conductance and thermopower in the tunneling regime. Operation in the tunneling regime would enable dramatic enhancements in the spatial resolution of more than two orders of magnitude, allowing experiments to probe longstanding theoretical predictions on quantum coherent conductors. The proposed method relies on standard scanning tunneling microscopy (STM) techniques and we describe the operation in detail in the subsequent sections. Our method is applicable to any nanoscopic conductor where the Wiedemann–Franz (WF) law holds quite generally [25].

The WF law is the statement of a rather simple observation about heat transport in electrical conductors. It asserts that in electrical conductors, heat is carried mainly by the electrons and that therefore electrical (σ) and thermal conductance (κ) must be related in a material-independent way $\kappa = \sigma L T$, where $L = \pi^2 k_B^2 / 3e^2$ is a universal constant called the Lorenz number and T is the temperature. While its applicability in bulk metals has been known for over 150 years, it can be shown to hold true in quantum systems where transport is dominated by elastic processes (see Appendix C, Sect. C.2, for the derivation). Recently, there have been experimental verifications of the WF law in atomic contact junctions [26, 27].

Scanning Tunneling Thermometer

From a fundamental point of view, a thermometer is a device that equilibrates locally with the system of interest and has some temperature-dependent physical property (e.g., resistance, thermopower, mass density) which can be measured; the temperature measurement seeks to find the condition(s) under which the thermometer is in local thermodynamic equilibrium with the system of interest and concurrently infers the thermometer's temperature by relying upon those temperature-dependent physical properties. Ideally, the measurement apparatus must not substantially disturb the state of the system of interest. SThM schemes, by relying on heat fluxes in the contact regime, may alter the state of a small system. We propose here a noninvasive thermometer whose local equilibration can be inferred by the measurement of electrical tunneling currents alone. In particular, we find that the conditions required for the local equilibration of the *scanning tunneling thermometer* (STT) are completely determined by (a) the conductance and thermopower which are both measured using the tunneling current and (b) the known bias conditions of the conductor defined by the voltages and temperatures of the contacts.

4.2 Temperature Measurement

In operating nanoscale devices, it is quite clear that the voltage refers to the electronic voltage since they are the only charged species participating in the transport. However, it is less clear what is meant by a temperature in this context since heat can be carried by other degrees of freedom such as phonons and photons as well. Out of equilibrium, the temperatures of the different degrees of freedom do not coincide. A majority of nanoscale electronic devices operate in the elastic transport regime where electron and phonon degrees of freedom are completely decoupled, and the distinction between their temperatures become extremely important. SThM measurements cannot make this distinction since they rely on measuring heat-flux-related signals which carry contributions from all degrees of freedom. *The method proposed here provides* the much needed measurement of *the electronic temperature, decoupled from all other degrees of freedom,* and can therefore also be used as a tool to characterize nonequilibrium device performance. In the remainder of the chapter, we refer only to the electronic temperature.

4.2.1 An Important Note

We note a crucial, but often overlooked, theoretical point pertaining to the imaging of temperature fields on a nonequilibrium conductor. The prevailing paradigm for temperature and voltage measurements is the following [28]: (1) a voltage is measured by a probe (voltmeter/thermometer) when in electrical equilibrium and

(2) a temperature is measured when in thermal equilibrium with the sample. We refer to this definition as the Engquist–Anderson (EA) definition. The fact that the EA definition implicitly ignores thermoelectric effects was pointed out by Bergfield and Stafford [20], and a notion of a joint probe was put forth by requiring *both* electrical and thermal equilibrium with the sample. Physically, one can think of it in the following manner. A temperature probe which is not in electrical equilibrium with the sample can develop a temperature bias at the probe-sample junction due to the Peltier effect. Similarly, a voltage probe which is not in thermal equilibrium with the sample can develop a voltage bias at the probe-sample junction due to the Seebeck effect. These errors can be quite large in quantum coherent conductors as reported in Ref. [29]. A temperature probe therefore has to remain in thermal *and* electrical equilibrium with the nonequilibrium sample [20, 21, 29–32], thereby ensuring true thermodynamic equilibrium of the measurement apparatus with the nonequilibrium sample.

This joint probe measurement was made mathematically rigorous in Chap. 2 where it was shown that the probe equilibration always exists and is unique, arbitrarily far from equilibrium and with arbitrary interactions within the quantum system. Moreover, we showed that the EA definition is provably nonunique: The value measured by the EA thermometer depends quite strongly on its voltage and, conversely, the value measured by the EA voltmeter depends on its temperature. We quickly recapitulate these results: Lemma 2.1 (Lemma 2.2) shows that a voltage (temperature) is uniquely defined only when the temperature (voltage) is specified. Theorem 2.2 showed that both the temperature and voltage are uniquely defined if they are measured simultaneously. These results are intimately tied to the second law of thermodynamics and expose the inconsistency in the EA definition. We need to treat temperature and voltage measurements on an equal footing. A measurement of *electron* temperature therefore has to involve also a simultaneous measurement of voltage simply due to the fact that electrons carry both charge and heat. Therefore we may write

$$I_p = J_p = 0, \tag{4.1}$$

for the simultaneous vanishing of the electrical current[1] I_p and the *electronic contribution* to the heat current J_p.

4.3   Temperature from Tunneling Currents

The scanning tunneling thermometer involves an STM tip scanning the surface of the conductor at a constant height in the tunneling regime as shown in Fig. 4.1a. We

---

[1]In a previous version of this dissertation, we used the electron particle current instead of the electrical current. These are equivalent conditions but result in an additional factor of e in defining the linear response coefficient $\mathcal{L}_{p\alpha}^{(0)}$ in Eq. (4.2).

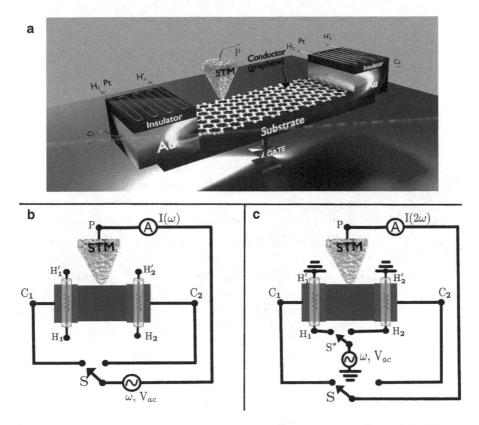

Fig. 4.1 (a) Schematic depiction of the system and measurement apparatus [1]. An STM tip scans the surface of a nanoscale conductor at a fixed height. The conductor sits atop a substrate which may be gated. Two gold contacts C_1 and C_2 are connected to the conductor on either side. A Pt heater ($H_1 H_1'$ and $H_2 H_2'$) sitting atop each gold contact, and electrically insulated from it, allows one to modulate the temperature of the gold contacts. (b) The conductance circuit which measures the coefficient $\mathcal{L}_{p\alpha}^{(0)}$ for each contact α selected using switch S. (c) The thermoelectric circuit which measures the coefficient $\mathcal{L}_{p\alpha}^{(1)}$ for each contact α. Switch S* activates the heater in the corresponding contact α selected by switch S

propose to measure simultaneously the conductance and thermopower by inducing time-modulated changes in the probe-contact voltage and the contact temperature, respectively. Our method is fully compatible with standard lock-in techniques so as to overcome the noise due thermal fluctuations at finite operating temperatures. The circuit operations for the measurement of conductance and thermopower are shown in Fig. 4.1b and c respectively and have been described in Sect. 4.4. Although simultaneous measurements of the conductance and thermopower have been made previously (e.g., in Ref. [33] on single-molecule break junctions), their relevance as a joint probe for temperature and voltage in nonequilibrium systems has remained entirely unappreciated. In this section, we relate the temperature measurement to the conductance and thermopower and, in the subsequent Sect. 4.4, we describe its experimental implementation.

Within the linear response regime, the probe currents depend linearly on the temperature and voltage gradients

$$\begin{pmatrix} I_p \\ J_p \end{pmatrix} = \sum_{\alpha} \begin{pmatrix} \mathcal{L}_{p\alpha}^{(0)} & \mathcal{L}_{p\alpha}^{(1)} \\ \mathcal{L}_{p\alpha}^{(1)} & \mathcal{L}_{p\alpha}^{(2)} \end{pmatrix} \begin{pmatrix} V_{\alpha} - V_p \\ \frac{T_{\alpha} - T_p}{T_0} \end{pmatrix}, \tag{4.2}$$

where $\mathcal{L}_{p\alpha}^{(\nu)}$ are the Onsager linear response coefficients evaluated at the equilibrium temperature T_0 and chemical potential μ_0. They have been explicitly defined in Appendix C.[2] $\mathcal{L}_{p\alpha}^{(0)}$ is the electrical conductance ($\mathcal{L}_{p\alpha}^{(0)} = G_{p\alpha}$) between the probe p and contact α. $\mathcal{L}_{p\alpha}^{(1)}$ is related to the thermopower ($S_{p\alpha}$) and electrical conductance ($\mathcal{L}_{p\alpha}^{(1)} = -T_0 S_{p\alpha} G_{p\alpha}$). Finally, $\mathcal{L}_{p\alpha}^{(2)}$ is related to the thermal conductance ($\mathcal{L}_{p\alpha}^{(2)} = T_0 \kappa_{p\alpha}$) [up to leading order in the Sommerfeld series [25] (see Appendix C)].

Equation (4.2) suggests that $\mathcal{L}_{p\alpha}^{(0)}$ and $\mathcal{L}_{p\alpha}^{(1)}$ can be measured using the tunneling current I_p, whereas $\mathcal{L}_{p\alpha}^{(2)}$ appears only in the expression for the heat current J_p and would generally involve the measurement of a heat-flux-related signal. However, for systems obeying the WF law,[3] we may simply related the $\mathcal{L}_{p\alpha}^{(0)}$ and $\mathcal{L}_{p\alpha}^{(2)}$ by

$$\mathcal{L}_{p\alpha}^{(2)} = \frac{\pi^2 k_B^2 T_0^2}{3e^2} \mathcal{L}_{p\alpha}^{(0)}, \tag{4.3}$$

to leading order in the Sommerfeld series.

We may write down the exact solution to Eq. (4.1) in the linear response regime [21]

$$\frac{T_p^{(\text{Exact})}}{T_0} = \frac{\sum_{\beta} \mathcal{L}_{p\beta}^{(0)} \sum_{\alpha} \mathcal{L}_{p\alpha}^{(1)} V_{\alpha} - \sum_{\beta} \mathcal{L}_{p\beta}^{(1)} \sum_{\alpha} \mathcal{L}_{p\alpha}^{(0)} V_{\alpha}}{\sum_{\beta} \mathcal{L}_{p\beta}^{(2)} \sum_{\alpha} \mathcal{L}_{p\alpha}^{(0)} - \left(\sum_{\alpha} \mathcal{L}_{p\alpha}^{(1)}\right)^2}$$
$$+ \frac{1}{T_0} \frac{\sum_{\beta} \mathcal{L}_{p\beta}^{(0)} \sum_{\alpha} \mathcal{L}_{p\alpha}^{(2)} T_{\alpha} - \sum_{\beta} \mathcal{L}_{p\beta}^{(1)} \sum_{\alpha} \mathcal{L}_{p\alpha}^{(1)} T_{\alpha}}{\sum_{\beta} \mathcal{L}_{p\beta}^{(2)} \sum_{\alpha} \mathcal{L}_{p\alpha}^{(0)} - \left(\sum_{\alpha} \mathcal{L}_{p\alpha}^{(1)}\right)^2}. \tag{4.4}$$

Here $T_p^{(\text{Exact})}$ denotes the exact solution to the equilibration of the STT, i.e., Eq. (4.1), within the linear response regime where the currents are expressed by Eq. (4.2).

---

[2] $\mathcal{L}_{p\alpha}^{(\nu)}$ do not coincide with definitions used in Refs. [31, 32, 34] and Chap. 2: e.g., $\mathcal{L}_{p\alpha}^{(0)}$ used here has an additional factor of e^2. One factor of e appears since we use electrical current instead of the particle current (see footnote 1). An additional factor of e appears since we write the currents in terms of the bias voltage instead of the bias chemical potential in Eq. (4.2).

[3] The Wiedemann–Franz law can be derived explicitly for elastic transport as shown in Appendix C. The first two terms in the Sommerfeld series are shown explicitly.

Our proposed method relies on the WF law given by Eq. (4.3), and to leading order in the Sommerfeld expansion we have

$$\frac{T_p^{(WF)}}{T_0} = \frac{3}{\pi^2 k_B^2 T_0^2} \left(\frac{\sum_\alpha \mathcal{L}_{p\alpha}^{(1)} \mu_\alpha}{\sum_\alpha \mathcal{L}_{p\alpha}^{(0)}} - \frac{\sum_\alpha \mathcal{L}_{p\alpha}^{(1)}}{\sum_\alpha \mathcal{L}_{p\alpha}^{(0)}} \frac{\sum_\beta \mathcal{L}_{p\beta}^{(0)} \mu_\beta}{\sum_\beta \mathcal{L}_{p\beta}^{(0)}} \right) + \frac{\sum_\alpha \mathcal{L}_{p\alpha}^{(0)} T_\alpha}{T_0 \sum_\alpha \mathcal{L}_{p\alpha}^{(0)}}.$$

(4.5)

$T_p^{(WF)}$ requires only the measurement of $\mathcal{L}_{p\alpha}^{(0)}$ and $\mathcal{L}_{p\alpha}^{(1)}$ and lends itself to a simple interpretation. The first term on the r.h.s. in Eq. (4.5) is the thermoelectric contribution whereas the second term is the thermal contribution. The second-order corrections (see Appendix C for more details) in the Sommerfeld series are typically very small

$$T_p^{(WF)} = T_p^{(Exact)} \left(1 + \mathcal{O}\left(\frac{k_B T_0}{\Delta} \right)^2 + \cdots \right),$$

(4.6)

where the characteristic energy scale of the problem Δ is typically much larger than the thermal energy set by $k_B T_0$: e.g., $\Delta = \epsilon_F$, the Fermi energy, for bulk systems and for a tunneling probe Δ is of the order of the work function.

It is clear from Eq. (4.5) that the measurement of (a) conductance $\mathcal{L}_{p\alpha}^{(0)}$ and the thermoelectric coefficient $\mathcal{L}_{p\alpha}^{(1)}$ along with the (b) known bias conditions of the system $\{V_\alpha, T_\alpha\}$ completely determines the conditions under which the STT is in local thermodynamic equilibrium with the system. The simultaneous measurement of the electrical conductance and thermopower therefore determines both the local temperature and voltage.

A Note on the Validity of the WF Law The breakdown of the Wiedemann–Franz law has been reported in various nanoscale systems. The characteristic energy scale Δ in such cases is comparable to the thermal energy thereby leading to large corrections from the higher-order terms in the series expansion (4.6). In graphene, the breakdown of the WF law was reported in Ref. [35]. Here, the local chemical potential was tuned (via local doping) such that it is smaller than the thermal energy thereby creating the so-called Dirac fluid. Such systems show a decoupling of charge and heat currents, making it impossible to measure heat currents through electrical means. Although our results apply to a broad array of nanoscale conductors, they do not apply to systems prepared in this manner.

4.4 Experimental Implementation

The temperature measurement involves two circuits: (I) The conductance circuit which measures the electrical conductance $\mathcal{L}_{p\alpha}^{(0)}$ and (II) The thermoelectric circuit which measures the thermoelectric response coefficient $\mathcal{L}_{p\alpha}^{(1)}$, as shown in Fig. 4.1b and c respectively. The STT involves operating the tip of a scanning tunneling microscope (STM) at a constant height above the surface of the conductor in the tunneling regime. The circuit operations (I) and (II) are described below.

(I) *The conductance circuit* involves a closed circuit of the probe and the contact α. All contacts and the probe are held at the equilibrium temperature $T_\alpha = T_p = T_0$. An AC voltage $V(\omega)$ is applied at the probe-contact junction $V(\omega) = V_p - V_\alpha$ and the resulting tunneling current $I_p(\omega)$ is recorded using standard lock-in techniques. The STM tip is scanned along the surface. A switch disconnects all contacts except α and the tunneling current is therefore

$$I_p = \mathcal{L}_{p\alpha}^{(0)}(V_p - V_\alpha) = -I_\alpha,$$
$$I_p(\omega) = \mathcal{L}_{p\alpha}^{(0)} V(\omega). \tag{4.7}$$

The procedure is repeated for all the contacts α by toggling the switch S shown in Fig. 4.1b and a scan is obtained for each probe-contact junction. This completes the measurement of the conductance $\mathcal{L}_{p\alpha}^{(0)}$ for all the contacts α.

(II) *The thermoelectric circuit* involves a (1) closed circuit of the probe and contact α, which is the same as the conductance circuit without the voltage source, and (2) an additional circuit which induces time-modulated temperature variations in contact α; An AC current at frequency ω induces Joule heating in the Pt resistor at frequency 2ω and results in a temperature modulation $T_\alpha = T_0 + \Delta T_\alpha(2\omega)$ in the contact α. The probe is held at the equilibrium temperature $T_p = T_0$. The resulting tunneling current $I_p(2\omega)$, at frequency 2ω, is recorded using standard lock-in techniques. The STM tip is scanned along the surface at the same points as before. A switch disconnects all contacts except α and the tunneling current is

$$I_p = \mathcal{L}_{p\alpha}^{(1)} \frac{(T_\alpha - T_p)}{T_0} = -I_\alpha$$
$$I_p(2\omega) = \mathcal{L}_{p\alpha}^{(1)} \frac{\Delta T_\alpha(2\omega)}{T_0}. \tag{4.8}$$

The procedure is repeated for all the contacts α by toggling the switches S and S* shown in Fig. 4.1c and a scan is obtained for each probe-contact junction. Note that the switch S* must heat the Pt resistor in the same contact α for which the probe-contact tunneling current is measured. This completes the measurement of the thermoelectric coefficient $\mathcal{L}_{p\alpha}^{(1)}$ for all the contacts α.

The voltage modulation frequency in the heating elements $\omega \ll 1/\tau$, where τ is the thermal time constant of the contacts, so that the contact may thermalize with the heating element. Typically, τ is of the order of tens of nanoseconds (cf. methods in [20]). We discuss the calibration of the contact temperature $T_\alpha = T_0 + \Delta T_\alpha$ in Appendix D. The thermoelectric response of the nanosystem may be quite sensitive to the gate voltage which is discussed in Appendix D.

Heating elements have been fabricated in the contacts previously [36]. Any system where one may induce Joule heating can be used as the heating element in the circuit. For example, another flake of graphene could be used as a heating element as long as it is calibrated accurately. We discuss the calibration of the temperature in the contacts in Appendix D.

4.5 Numerical Results and Discussion

We present model temperature measurements for a hexagonal graphene flake under (a) a thermal bias and (b) a voltage bias. The measured temperature, for a combination of thermal and voltage biases, would simply be a linear combination of the two scenarios (a) and (b) in the linear response regime (under identical gating conditions). Therefore, we present the two cases separately but we note that the gate voltages are not the same for the two scenarios that we present here. The voltage bias case has been gated differently so as to enhance the thermoelectric response of the system (see Appendix D for a more detailed discussion). We show the temperature measurement for (a) the thermal bias case in Fig. 4.2 and (b) the voltage bias case in Fig. 4.3. The two panels in Figs. 4.2 and 4.3 compare (1) the temperature measurement $T_p^{(\text{Exact})}$ obtained from the exact solution [given by Eq. (4.4)] and (2) the temperature measurement $T_p^{(\text{WF})}$ obtained from our method [given by Eq. (4.5)] which relies on the WF law.

Graphene is highly relevant for future electronic technologies and provides a versatile system whose transport properties can be tuned by an appropriate choice of the gate voltage—we therefore illustrate our results for graphene. The method itself is valid for any system obeying the WF law. The thermoelectric response coefficient $\mathcal{L}_{p\alpha}^{(1)} \sim T_0^2$ has a quadratic suppression at low temperatures and its measurement from Eq. (4.8) depends crucially on the choice of gating especially at cryogenic operating temperatures ~ 4 K since the resulting tunneling current must be experimentally resolvable. In graphene, we find that the electrical tunneling currents arising from its thermoelectric response are resolvable even at cryogenic temperatures when the system is gated appropriately and, owing to the fact that a number of STM experiments are conducted at low temperatures, we present our results for $T_0 = 4$ K. Higher operating temperatures result in a higher tunneling current in Eq. (4.8) and gating would therefore be less important.

The π-electron system of graphene is described using the tight-binding model whose basis states are $2p_z$ orbitals at each atomic site of carbon. The STT is modeled as an atomically sharp Pt tip operating at a constant height of 3 Å above the plane of the carbon nuclei. The details of the graphene Hamiltonian as well as the probe-system tunnel coupling are presented in Sects. 4.6.1 and 4.6.2 respectively. The atomic sites of graphene which are coupled to the contacts are indicated in Figs. 4.2 and 4.3 by either a red or blue square. The chemical potential and temperature of the two contacts (red and blue) set the bias conditions for the problem. The coupling to

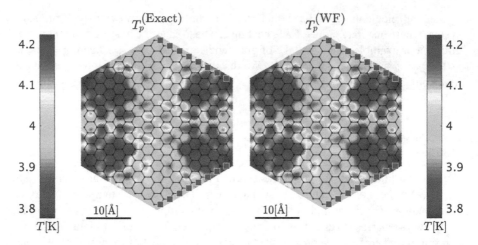

Fig. 4.2 Temperature variations on a hexagonal graphene flake with an application of a symmetrical temperature bias $T_{\text{red}} - T_{\text{blue}} = 0.5$ K, where the red (hot) and blue (cold) squares indicate the sites coupled to the contacts; $T_0 = 4$ K. The left panel shows the exact linear response solution given in Eq. (4.4) while the right panel shows the WF solution given by Eq. (4.5). The same temperature scale is used for both panels

the two contacts is symmetrical and the coupling strength for all the coupling sites (red or blue) is taken as $\Gamma = 0.5$ eV. Additional details regarding the gating and the tunneling currents are given in Appendix D.

Figure 4.2 shows the variations in temperature for a symmetrical ($T_{\text{red}} + T_{\text{blue}} = 2T_0$) temperature bias $T_{\text{red}} - T_{\text{blue}} = 0.5$ K. The agreement between $T_p^{\text{(Exact)}}$ and $T_p^{\text{(WF)}}$ given by Eqs. (4.4) and (4.5) respectively is excellent. The gating has been chosen to be $\mu_0 = -2.58$ eV with respect to the Dirac point in graphene. The same temperature scale is used for both the panels in Fig. 4.2. The temperature variations in $T_p^{\text{(WF)}}$ are solely the result of the temperature bias and are given by the second term in Eq. (4.5). Therefore, we require only the measurement of the conductances $\mathcal{L}_{p\alpha}^{(0)}$ for the temperature measurement under these bias conditions. We consider a contact-tip voltage modulation of 1 mV for the measurement of the conductance. The resulting tunneling currents are of the order of 10 nA with a maximum tunneling current of about 30 nA. We present the details in Appendix D.

Figure 4.3 shows the variations in temperature for a voltage bias of $V_{\text{blue}} - V_{\text{red}} = k_B T_0/e$, with $T_0 = 4$ K, so that the transport is within the linear response regime. The gating for this case has been chosen to be $\mu_0 = -2.28$ eV such that there is an enhanced thermoelectric effect. The tunneling currents from the thermoelectric circuit, under these gating conditions, are of the order of 100 pA with a maximum tunneling current of about $I = 150$ pA and are resolvable under standard lock-in techniques. The variation of the contact temperature is taken to be $\Delta T = (10\%) T_0$ with $T_0 = 4$ K. The resolution of the tunneling current is an important point especially for the measurement of the thermoelectric response coefficient $\mathcal{L}_{p\alpha}^{(1)}$ and

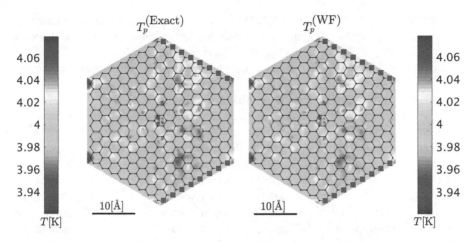

Fig. 4.3 Temperature variations on a hexagonal graphene flake with an application of a voltage bias $V_{\text{blue}} - V_{\text{red}} = k_B T_0 / e = 0.34\,\text{mV}$, where the red and blue squares indicate the sites coupled to the contacts; $T_0 = 4\,\text{K}$. The left panel shows the exact linear response solution given in Eq. (4.4) while the right panel shows the approximate solution obtained by employing the WF law given in Eq. (4.5). The same temperature scale is used for both panels

has been covered in greater detail in Appendix D. The same temperature scale is used for both panels in Fig. 4.3 and there is excellent agreement between $T_p^{(\text{Exact})}$ and $T_p^{(\text{WF})}$. The temperature variations shown here are solely the result of the voltage bias and are given by the first term in Eq. (4.5). Since the variations are purely due to the thermoelectric effect, the EA definition would have noted no temperature variations at all.

The disagreement between the exact solution and our method are due to higher-order contributions in the Sommerfeld series which are extremely small [cf. Eq. (4.6)]. An explicit expression for the first Sommerfeld correction in the WF law has been derived in Appendix C. The discrepancy between $T_p^{(\text{WF})}$ and $T_p^{(\text{Exact})}$ defined by $|T_p^{(\text{WF})} - T_p^{(\text{Exact})}| / T_p^{(\text{Exact})}$ is less than 0.01% for the temperature bias case in Fig. 4.2, whereas their discrepancy for the voltage bias case in Fig. 4.3 is less than 0.2%.

4.6 Theoretical Model: Additional Details

4.6.1 System Hamiltonian

The π-electron system of graphene is described within the tight-binding model, $H_{\text{gra}} = -\sum_{<i,j>} t_{ij} d_i^\dagger d_j + \text{h.c}$, with nearest-neighbor hopping matrix element $t_{ij} \equiv t = 2.7\,\text{eV}$. The coupling of the system with the contact reservoirs is described by the tunneling-width matrices Γ^α. We calculate the transport properties using nonequilibrium Green's functions. The retarded Green's function of the junction is given by $G^r(\omega) = [\mathbb{S}\omega - H_{\mathrm{gra}} - \Sigma_T(\omega)]^{-1}$, where $\Sigma_T = -i\sum_\alpha \Gamma^\alpha/2$ is the tunneling self-energy. We take the contact-system couplings in the broadband limit, i.e., $\Gamma^\alpha_{nm}(\omega) = \Gamma^\alpha_{nm}(\mu_0)$ where μ_0 is the Fermi energy of the metal leads. We also take the contact-system couplings to be diagonal matrices $\Gamma^\alpha_{nm}(\omega) = \sum_{l\in\alpha} \Gamma_\alpha \delta_{nl}\delta_{ml}$ coupled to π-orbitals n, m of the graphene system. The nonzero elements of Γ^α ($\alpha = \{$blue, red$\}$) are at sites indicated by either a blue or red square in Figs. 4.2 and 4.3, corresponding to the carbon atoms of graphene covalently bonded to the contact reservoirs. The tunneling matrix element at each coupling site is set as $\Gamma_\alpha = 0.5\,\mathrm{eV}$ for both the contacts (blue and red). $\mathbb{S}$ is the overlap-matrix between the atomic orbitals on different sites and we take $\mathbb{S} = \mathbb{1}$, i.e., an orthonormal set of atomic orbitals. The tunneling-width matrix Γ^p describing the probe-sample coupling is also treated in the broadband limit. The probe is in the tunneling regime and the probe-system coupling is weak (few meV) in comparison to the system-reservoir couplings.

4.6.2   Probe-Sample Coupling

The scanning tunneling thermometer is modeled as an atomically sharp Pt tip operating in the tunneling regime at a height of 3 Å above the plane of the carbon nuclei in graphene. The probe tunneling-width matrices may be described in general as [37] $\Gamma^p_{nm}(\omega) = 2\pi \sum_{l\in\{s,p,d...\}} C_l V^n_l V^{n*}_l \rho^p_l(\omega)$, where $\rho^p_l(\omega)$ is the local density of states of the apex atom in the probe electrode and V^m_l, V^n_l are the tunneling matrix elements between the l-orbital of the apex atom in the probe and the mth, nth π-orbitals in graphene. The constants $C_l = C \;\forall l$ and has been determined by matching with the peak of the experimental conductance histogram [38]. We consider the Pt tip to be dominated by the d-orbital character (80%) although other contributions (s—10% and p—10%) are also taken as described in Ref. [37]. In the calculation of the tunneling matrix elements, the π-orbitals of graphene are taken to be hydrogenic $2p_z$ orbitals with an effective nuclear charge $Z = 3.22$ [39]. The tunneling-width matrix Γ^p describing the probe-system coupling is in general non-diagonal.

4.7   Conclusions

It has proven extraordinarily challenging to achieve high spatial resolution in thermal measurements. A key obstacle has been the fundamental difficulty in designing a thermal probe that exchanges heat with the system of interest but is thermally isolated from the environment. We propose circumventing this seemingly intractable problem by inferring thermal signals using purely electrical measurements. The basis of our approach is the Wiedemann–Franz law relating the thermal and electrical currents flowing between a probe and the system of interest.

We illustrate this new approach to nanoscale thermometry with simulations of a scanning tunneling probe of a model nanostructure consisting of a graphene flake under thermoelectric bias. We show that the local temperature inferred from a sequence of purely electrical measurements agrees exceptionally well with that of a hypothetical thermometer coupled locally to the system and isolated from the environment. Moreover, our measurement provides the *electronic* temperature decoupled from all other degrees of freedom and can therefore be a vital tool to characterize nonequilibrium device performance. Our proposed *scanning tunneling thermometer* exceeds the spatial resolution of current state-of-the-art thermometry by some two orders of magnitude.

4.8   Note on Experimental Realization

In a previous version of our dissertation, we had proposed a different but equivalent measurement setting where we envisioned a heating element adjacent to the STM tip.[4] It was brought to our attention, following discussions with Oliver Monti and Brian LeRoy, that the fabrication of a calibrated thermoresistor on the STM tip is experimentally challenging. However, heating elements can be fabricated in the contact reservoirs which drive the system out of equilibrium, for example, as reported recently in Ref. [36]. It was further suggested to us that a flake of graphene can itself be used as a heating element since it undergoes Joule heating when sufficient current is passed. Therefore, a flake of graphene, along with a suitable substrate, can be used to induce the temperature modulations in the contact reservoirs as long as it can be calibrated properly. Furthermore, in the previous version of this dissertation, we presented our numerical results with the equilibrium temperature held at $T_0 = 100$ K. This is because we were concerned that the suppression of the thermoelectric effect at cryogenic temperatures would result in a tunneling current which is too small to resolve experimentally. However, further analysis suggested that the experiment could be performed at cryogenic temperatures as low as 4 K: Although the thermoelectric response is suppressed quadratically with decreasing temperature (cf. Sect. C.2 for the relevant Sommerfeld series expansion), it is possible to induce larger (experimentally resolvable) electrical currents by an appropriate gating of the system. We therefore have included an additional Appendix D where we discuss the details pertaining to the gating of the system and the resulting tunneling currents during the thermoelectric circuit operation.

---

[4]This would be an equivalent realization of our experiment since current conservation would imply that the transmission functions satisfy $\mathcal{T}_{p\alpha} = \mathcal{T}_{\alpha p}$. In principle, the experiment can be carried out even in the presence of a magnetic flux Φ but, since $\mathcal{T}_{p\alpha}(\Phi) = \mathcal{T}_{\alpha p}(-\Phi)$, we would have to invert the magnetic field to infer the $\mathcal{L}_{p\alpha}^{(\nu)}$ coefficients [40].

References

1. A. Shastry, S. Inui, C.A. Stafford, ArXiv e-prints 1901.09168 (2019)
2. G. Kucsko, P.C. Maurer, N.Y. Yao, M. Kubo, H.J. Noh, P.K. Lo, H. Park, M.D. Lukin, Nature **500**(7460), 54 (2013). http://dx.doi.org/10.1038/nature12373. Letter
3. C.Y. Jin, Z. Li, R.S. Williams, K.C. Lee, I. Park, Nano Lett. **11**(11), 4818 (2011). https://doi.org/10.1021/nl2026585. http://dx.doi.org/10.1021/nl2026585. PMID: 21967343
4. M. Mecklenburg, W.A. Hubbard, E.R. White, R. Dhall, S.B. Cronin, S. Aloni, B.C. Regan, Science **347**(6222), 629 (2015). https://doi.org/10.1126/science.aaa2433. http://science.sciencemag.org/content/347/6222/629
5. J.S. Reparaz, E. Chavez-Angel, M.R. Wagner, B. Graczykowski, J. Gomis-Bresco, F. Alzina, C.M.S. Torres, Rev. Sci. Instrum. **85**(3), 034901 (2014). https://doi.org/10.1063/1.4867166. http://dx.doi.org/10.1063/1.4867166
6. P. Neumann, I. Jakobi, F. Dolde, C. Burk, R. Reuter, G. Waldherr, J. Honert, T. Wolf, A. Brunner, J.H. Shim, D. Suter, H. Sumiya, J. Isoya, J. Wrachtrup, Nano Lett. **13**(6), 2738 (2013). https://doi.org/10.1021/nl401216y. http://dx.doi.org/10.1021/nl401216y. PMID: 23721106
7. D. Teyssieux, L. Thiery, B. Cretin, Rev. Sci. Instrum. **78**(3), 034902 (2007). https://doi.org/10.1063/1.2714040. http://dx.doi.org/10.1063/1.2714040
8. A. Majumdar, Annu. Rev. Mater. Sci. **29**, 505 (1999). https://doi.org/10.1146/annurev.matsci.29.1.505
9. K. Kim, W. Jeong, W. Lee, P. Reddy, ACS Nano **6**(5), 4248 (2012). https://doi.org/10.1021/nn300774n. http://dx.doi.org/10.1021/nn300774n. PMID: 22530657
10. W. Jeong, S. Hur, E. Meyhofer, P. Reddy, Nanoscale Microscale Thermophys. Eng. **19**(4), 279 (2015). https://doi.org/10.1080/15567265.2015.1109740. http://dx.doi.org/10.1080/15567265.2015.1109740
11. F. Menges, P. Mensch, H. Schmid, H. Riel, A. Stemmer, B. Gotsmann, Nat. Commun. **7**, 10874 EP (2016). http://dx.doi.org/10.1038/ncomms10874. Article
12. P. Muralt, D.W. Pohl, Appl. Phys. Lett. **48**(8), 514 (1986). https://doi.org/http://dx.doi.org/10.1063/1.96491. http://scitation.aip.org/content/aip/journal/apl/48/8/10.1063/1.96491
13. B.G. Briner, R.M. Feenstra, T.P. Chin, J.M. Woodall, Phys. Rev. B **54**, R5283 (1996). https://doi.org/10.1103/PhysRevB.54.R5283. https://link.aps.org/doi/10.1103/PhysRevB.54.R5283
14. G. Ramaswamy, A.K. Raychaudhuri, Appl. Phys. Lett. **75**(13), 1982 (1999). https://doi.org/10.1063/1.124892. http://dx.doi.org/10.1063/1.124892
15. W. Wang, K. Munakata, M. Rozler, M.R. Beasley, Phys. Rev. Lett. **110**(23) (2013). https://doi.org/10.1103/PhysRevLett.110.236802
16. K.W. Clark, X.G. Zhang, G. Gu, J. Park, G. He, R.M. Feenstra, A.P. Li, Phys. Rev. X **4**, 011021 (2014). https://doi.org/10.1103/PhysRevX.4.011021. http://link.aps.org/doi/10.1103/PhysRevX.4.011021
17. P. Willke, T. Druga, R.G. Ulbrich, M.A. Schneider, M. Wenderoth, Nat. Commun. **6**, 6399 (2015). http://dx.doi.org/10.1038/ncomms7399
18. R. Landauer, IBM J. Res. Dev. **1**(3), 223 (1957). https://doi.org/10.1147/rd.13.0223
19. Y. Dubi, M. Di Ventra, Nano Lett. **9**, 97 (2009)
20. J.P. Bergfield, S.M. Story, R.C. Stafford, C.A. Stafford, ACS Nano **7**(5), 4429 (2013). https://doi.org/10.1021/nn401027u
21. J. Meair, J.P. Bergfield, C.A. Stafford, P. Jacquod, Phys. Rev. B **90**, 035407 (2014). https://doi.org/10.1103/PhysRevB.90.035407. http://link.aps.org/doi/10.1103/PhysRevB.90.035407
22. J.P. Bergfield, M.A. Ratner, C.A. Stafford, M. Di Ventra, Phys. Rev. B **91**, 125407 (2015). https://doi.org/10.1103/PhysRevB.91.125407. http://link.aps.org/doi/10.1103/PhysRevB.91.125407
23. K. Kim, J. Chung, G. Hwang, O. Kwon, J.S. Lee, ACS Nano **5**(11), 8700 (2011). https://doi.org/10.1021/nn2026325. http://dx.doi.org/10.1021/nn2026325. PMID: 21999681

24. S. Gomès, A. Assy, P.O. Chapuis, Phys. Status Solidi A **212**(3), 477 (2015). https://doi.org/10.1002/pssa.201400360. http://dx.doi.org/10.1002/pssa.201400360
25. N.W. Ashcroft, N.D. Mermin, *Solid State Physics* (Brooks/Cole - Thomson Learning, Pacific Grove 1976)
26. N. Mosso, U. Drechsler, F. Menges, P. Nirmalraj, S. Karg, H. Riel, B. Gotsmann, Nat Nano **12**(5), 430 (2017). Letter. http://dx.doi.org/10.1038/nnano.2016.302
27. L. Cui, W. Jeong, S. Hur, M. Matt, J.C. Klöckner, F. Pauly, P. Nielaba, J.C. Cuevas, E. Meyhofer, P. Reddy, Science **355**(6330), 1192 (2017). https://doi.org/10.1126/science.aam6622 http://science.sciencemag.org/content/355/6330/1192
28. H.L. Engquist, P.W. Anderson, Phys. Rev. B **24**, 1151 (1981). https://doi.org/10.1103/PhysRevB.24.1151. http://link.aps.org/doi/10.1103/PhysRevB.24.1151
29. J.P. Bergfield, C.A. Stafford, Phys. Rev. B **90**, 235438 (2014). https://doi.org/10.1103/PhysRevB.90.235438. http://link.aps.org/doi/10.1103/PhysRevB.90.235438
30. A. Shastry, C.A. Stafford, Phys. Rev. B **92**, 245417 (2015). https://doi.org/10.1103/PhysRevB.92.245417. http://link.aps.org/doi/10.1103/PhysRevB.92.245417
31. C.A. Stafford, Phys. Rev. B **93**, 245403 (2016). https://doi.org/10.1103/PhysRevB.93.245403. http://link.aps.org/doi/10.1103/PhysRevB.93.245403
32. A. Shastry, C.A. Stafford, Phys. Rev. B **94**, 155433 (2016). https://doi.org/10.1103/PhysRevB.94.155433. http://link.aps.org/doi/10.1103/PhysRevB.94.155433
33. J.R. Widawsky, P. Darancet, J.B. Neaton, L. Venkataraman, Nano Lett. **12**(1), 354 (2012). https://doi.org/10.1021/nl203634m. http://dx.doi.org/10.1021/nl203634m. PMID: 22128800
34. J.P. Bergfield, C.A. Stafford, Nano Lett. **9**, 3072 (2009)
35. J. Crossno, J.K. Shi, K. Wang, X. Liu, A. Harzheim, A. Lucas, S. Sachdev, P. Kim, T. Taniguchi, K. Watanabe, T.A. Ohki, K.C. Fong, Science **351**(6277), 1058 (2016). https://doi.org/10.1126/science.aad0343. http://science.sciencemag.org/content/351/6277/1058
36. M. Tsutsui, T. Kawai, M. Taniguchi, Sci. Rep. **2** (2012). Article Number 21. http://dx.doi.org/10.1038/srep00217
37. J.P. Bergfield, J.D. Barr, C.A. Stafford, Beilstein J. Nanotechnol. **3**, 40 (2012). https://doi.org/10.3762/bjnano.3.5
38. M. Kiguchi, O. Tal, S. Wohlthat, F. Pauly, M. Krieger, D. Djukic, J.C. Cuevas, J.M. van Ruitenbeek, Phys. Rev. Lett. **101**, 046801 (2008)
39. J.D. Barr, C.A. Stafford, J.P. Bergfield, Phys. Rev. B **86**, 115403 (2012). https://doi.org/10.1103/PhysRevB.86.115403. https://link.aps.org/doi/10.1103/PhysRevB.86.115403
40. M. Büttiker, Phys. Rev. Lett. **57**, 1761 (1986)

Chapter 5
Entropy

In this chapter, we shall discuss entropy in the context of the steady-state quantum transport problem. The work presented here deals with noninteracting fermions. We consider situations where we have a few reservoirs which exchange particles and energy with each other through a central scattering region but the reservoirs themselves are taken to be in the thermodynamic limit. This means that the exchange of particles and energy does not alter the distribution within the reservoirs. The distribution within the reservoirs, specified by their temperature and chemical potential, set the boundary conditions for the scattering problem. Before proceeding to the discussion of the scattering problem, we first give a basic overview of entropy in quantum mechanics in Sect. 5.1.

In Sect. 5.2 we formulate the entropy in terms of the scattering states and find that, in the scattering basis, the total entropy is additive over the different reservoirs since their corresponding scattering states do not mix in the absence of two-body interactions. We then formulate the entropy that is measured by a local observer who does not possess detailed knowledge of the scattering states but has detailed knowledge of the local spectrum and local distribution function. We find that this lack of knowledge increases the observed entropy. Finally, we formulate the entropy that is measured without knowledge of the local distribution function but having knowledge of the local mean particle number and energy. We show that the entropies formulated in the three different ways satisfy a hierarchy of inequalities (Theorems 5.1 and 5.2) with the most knowledgable formulation leading to the least entropy. We illustrate our results for a two-level system far from equilibrium. A detailed discussion of the third law of thermodynamics in open quantum systems is also included. We show (see Theorem 5.3 and Corollary 5.3.1) that fully quantum open systems (having no localized states) will have vanishing entropy at zero temperature whereas generic open quantum systems with localized states may, very rarely, have finite contributions due to such localized states. The contribution due to such localized states is shown to be $g \log(2)$ where g is the number of localized states in the system lying exactly at the chemical potential of

© Springer Nature Switzerland AG 2019
A. Shastry, *Theory of Thermodynamic Measurements of Quantum Systems Far from Equilibrium*, Springer Theses, https://doi.org/10.1007/978-3-030-33574-8_5

the reservoir. We then propose an ansatz for the entropy when two-body interactions are present. Based on the inequalities developed, we also use the entropy as a metric to quantify the "distance" from equilibrium by normalizing it appropriately. We also show numerical results for a molecular junction driven far from equilibrium.

5.1   Preliminaries

Entropy is one of the most fundamental quantities in all of physics. The entropy for any quantum system is given by[1]

$$S = -\operatorname{Tr}\{\hat{\rho}\log\hat{\rho}\}. \tag{5.1}$$

Here $\hat{\rho}$ is the density operator describing the quantum system. The log operation is understood in much the same way we understand exponentials of operators. For example, we may write $e^{\hat{A}} = \hat{B}$ and understand that the l.h.s. is a series expansion of operator $\hat{A}$ which converges to $\hat{B}$. Then $\hat{A} = \log\hat{B}$ and is uniquely defined for any density operator. The expression for entropy in the form above is due to von Neumann [1] and is referred to as the von Neumann entropy.

Any density operator can be written in its complete many-body eigenbasis as $\hat{\rho} = \sum_i p_i |\Psi_i\rangle\langle\Psi_i|$ and the entropy is then easily calculated to be

$$S = -\sum_i p_i \log p_i, \tag{5.2}$$

which is a form often encountered in information theory. The p_i are the probabilities associated with states $|\Psi_i\rangle$. The density operator has unit trace $\sum_i p_i = 1$. The density operator of a system completely encodes all the statistical information pertaining to the system and is the central quantity in quantum statistical mechanics. The statistical average of any operator $\hat{O}$ is $\langle\hat{O}\rangle = \operatorname{Tr}\{\hat{\rho}\hat{O}\}$. One may therefore introduce the entropy operator as

$$\hat{S} = -\log\hat{\rho}, \tag{5.3}$$

so that the entropy in Eq. (5.1) is the statistical average of the above operator.

It is worth noting that entropy is invariant under unitary transformations of the density operator $\hat{\rho} \to \hat{U}^\dagger\hat{\rho}\hat{U}$. Let us take $\log\hat{\rho} = -\hat{S} = \hat{A}$ so that $\hat{\rho} = e^{\hat{A}}$. Then

---

[1]It is usual to include a prefactor of k_B which we here set to unity.

$$\hat{U}^\dagger \hat{\rho}\, \hat{U} = \hat{U}^\dagger (1 + \hat{A} + \frac{1}{2!}\hat{A}^2 + \ldots)\hat{U}$$

$$= 1 + \hat{U}^\dagger \hat{A}\hat{U} + \frac{1}{2!}\hat{U}^\dagger \hat{A}\hat{U}\, \hat{U}^\dagger \hat{A}\hat{U} + \ldots \qquad (5.4)$$

$$= \exp\left(\hat{U}^\dagger \hat{A}\hat{U}\right).$$

This implies that $\log\left(\hat{U}^\dagger \hat{\rho}\, \hat{U}\right) = \hat{U}^\dagger \log \hat{\rho}\, \hat{U}$ and the entropy

$$S = - \operatorname{Tr}\left\{\hat{U}^\dagger \hat{\rho}\hat{U} \log\left(\hat{U}^\dagger \hat{\rho}\hat{U}\right)\right\}$$

$$= - \operatorname{Tr}\left\{\hat{U}^\dagger \hat{\rho}\hat{U}\, \hat{U}^\dagger \log \hat{\rho}\, \hat{U}\right\} \qquad (5.5)$$

$$= - \operatorname{Tr}\left\{\hat{\rho} \log \hat{\rho}\right\}.$$

This means that a closed quantum system evolving under the Schrödinger equation will always have the same entropy. It is only measurements or interactions with an environment which can bring about a change in the system's entropy.

5.1.1 Single Fermionic Level

We start our discussion by writing down the density operator of a single fermionic level. We do this in the full many-body Hilbert space (also referred to as Fock space). Of course, the full many-body Hilbert space for a single Fermi level has only two linearly independent state vectors available to it. This would be the vacuum and the one-particle level. The Pauli principle restricts us from having more than one fermion in the same state. Nevertheless, this problem is instructive and can be easily generalized to include more states.

In general, the many-body Hilbert space that is accessible to systems of identical particles is a small subspace of the entire many-body Hilbert space. This is because symmetry constrains the state vectors that are accessible to identical particles obeying a symmetry principle. The allowed state vectors are either symmetric (bosons) or antisymmetric (fermions) with respect to exchange of any two particles. The natural framework to describe a system of identical particles is called second quantization. In the second quantization framework, we do not have to worry about the symmetry principles for the state vectors themselves but rather assume that the entire many-body Hilbert space accessible to them. The symmetry principle is then relegated to the creation and destruction operators which obey a certain commutation relation depending upon the type of symmetry (Fermi or Bose). This leads to a great simplification. The reader may refer to any standard textbook on quantum mechanics for an exposition to this subject.

In equilibrium, the density operator is written as

$$\hat{\rho} = \frac{e^{-\beta(\hat{H}-\mu\hat{N})}}{\text{Tr}\left\{e^{-\beta(\hat{H}-\mu\hat{N})}\right\}}, \tag{5.6}$$

where β is the inverse temperature and μ is the chemical potential. The single level Hamiltonian and number operators are

$$\hat{H} = \epsilon\hat{\Psi}^{\dagger}\hat{\Psi}$$
$$\hat{N} = \hat{\Psi}^{\dagger}\hat{\Psi}, \tag{5.7}$$

where $\hat{\Psi}^{\dagger}$ and $\hat{\Psi}$ are the creation and annihilation operators, respectively, for the single level with energy ϵ. We write the operator $\hat{Z}$ as

$$\hat{Z} = e^{-\beta(\hat{H}-\mu\hat{N})},$$
$$Z = \text{Tr}\left\{\hat{Z}\right\}. \tag{5.8}$$

The entropy operator [cf. Ref. [2, Eq. (2)]] then can be written as

$$\hat{S} = -\log\hat{\rho} = \beta(\epsilon - \mu)\hat{\Psi}^{\dagger}\hat{\Psi} + \log(Z), \tag{5.9}$$

and

$$-\hat{\rho}\,\log\hat{\rho} = \frac{\beta(\epsilon - \mu)}{Z}e^{-\beta(\epsilon-\mu)\hat{\Psi}^{\dagger}\hat{\Psi}}\hat{\Psi}^{\dagger}\hat{\Psi} + \frac{\log(Z)}{Z}e^{-\beta(\epsilon-\mu)\hat{\Psi}^{\dagger}\hat{\Psi}}. \tag{5.10}$$

The operator $\hat{Z}$ has the series expansion

$$\hat{Z} = \exp\left\{\left(-\beta(\epsilon - \mu)\hat{\Psi}^{\dagger}\hat{\Psi}\right)\right\}$$
$$= 1 - \beta(\epsilon - \mu)\hat{\Psi}^{\dagger}\hat{\Psi} + \frac{\beta^2(\epsilon - \mu)^2}{2!}\hat{\Psi}^{\dagger}\hat{\Psi}\hat{\Psi}^{\dagger}\hat{\Psi} + \ldots$$
$$= 1 + \hat{\Psi}^{\dagger}\hat{\Psi}\left(-1 + 1 - \beta(\epsilon - \mu) + \frac{\beta^2(\epsilon - \mu)^2}{2!} + \ldots\right) \tag{5.11}$$
$$= 1 + \hat{\Psi}^{\dagger}\hat{\Psi}(e^{-\beta(\epsilon-\mu)} - 1).$$

Here we have used the fact that $(\hat{\Psi}^{\dagger}\hat{\Psi})^2 = \hat{\Psi}^{\dagger}\hat{\Psi}$ for fermions. In the third line, we have added and subtracted 1 to obtain the exponential series. Note also that the first term in the last line is 1 and its action on state vectors is that of the identity operator on the many-body Hilbert space. Therefore, when we take the trace, the first term contributes twice (once for the vacuum and once for the single Fermi level ϵ):

$$Z = \text{Tr}\{\hat{Z}\} = 1 + 1 + (e^{-\beta(\epsilon-\mu)} - 1) = 1 + e^{-\beta(\epsilon-\mu)}. \tag{5.12}$$

The entropy can be calculated by taking the trace of Eq. (5.10)

$$\begin{aligned} S &= \frac{\beta(\epsilon-\mu)e^{-\beta(\epsilon-\mu)}}{1 + e^{-\beta(\epsilon-\mu)}} + \log Z \\ &= \beta(\epsilon-\mu)f(\epsilon) + \log\left(1 + e^{-\beta(\epsilon-\mu)}\right) \\ &= -\left(f\log f + (1-f)\log(1-f)\right), \end{aligned} \tag{5.13}$$

where a simple algebraic manipulation can show that lines 2 and 3 above are the same. The Fermi function gives the occupancy of the single Fermi level of energy ϵ:

$$f(\epsilon) = \langle\hat{\Psi}^\dagger\hat{\Psi}\rangle = \frac{1}{1 + e^{\beta(\epsilon-\mu)}}. \tag{5.14}$$

The argument presented above for the entropy can be made even more obvious by explicitly constructing the density matrix in the Hilbert space of a single fermionic level. The Hilbert space has exactly two orthonormal vectors which we label as $\{|0\rangle, |1\rangle\}$ corresponding to the vacuum state and the single-particle state of energy ϵ, respectively. It is easy to see that the density matrix given by Eq. (5.6) is diagonal in this basis and that

$$\hat{\rho} = \begin{pmatrix} \langle 0|\rho\rangle|0\rangle & \langle 0|\rho\rangle|1\rangle \\ \langle 1|\rho\rangle|0\rangle & \langle 1|\rho\rangle|1\rangle \end{pmatrix} = \begin{pmatrix} 1-f & 0 \\ 0 & f \end{pmatrix} = \begin{pmatrix} \frac{1}{Z} & 0 \\ 0 & \frac{e^{-\beta(\epsilon-\mu)}}{Z} \end{pmatrix}. \tag{5.15}$$

Therefore the probability associated with the vacuum is $1 - f$ while the probability associated with the level ϵ is f. The expressions for the entropy operator and entropy follow directly from the above expression. It is also straight-forward to calculate other average quantities.

In the second quantized notation, we may write Eq. (5.15) as

$$\hat{\rho} = \left((1-f)\hat{\Psi}\hat{\Psi}^\dagger + f\hat{\Psi}^\dagger\hat{\Psi}\right). \tag{5.16}$$

5.1.2 Generalization to Many States

It is straightforward to generalize the above analysis to a (noninteracting) system with multiple levels. The Hamiltonian and number operators are then given by

$$\hat{H} = \sum_i \hat{H}_i = \sum_i \epsilon_i \hat{\Psi}_i^\dagger \hat{\Psi}_i$$

$$\hat{N} = \sum_i \hat{N}_i = \sum_i \hat{\Psi}_i^\dagger \hat{\Psi}_i. \tag{5.17}$$

We note that the Hamiltonian and number operators corresponding to different levels always commute

$$[\hat{N}_i, \hat{N}_j] = 0. \tag{5.18}$$

Therefore the density operator can be written as

$$\hat{\rho} = \frac{e^{-\beta(\hat{H}-\mu\hat{N})}}{\mathrm{Tr}\left\{e^{-\beta(\hat{H}-\mu\hat{N})}\right\}} = \frac{\prod_i e^{-\beta(\epsilon_i-\mu)\hat{\Psi}_i^\dagger \hat{\Psi}_i}}{Z}, \tag{5.19}$$

where

$$Z = \mathrm{Tr}\left\{\hat{Z}\right\} = \mathrm{Tr}\left\{\Pi_i e^{-\beta(\epsilon_i-\mu)\hat{\Psi}_i^\dagger \hat{\Psi}_i}\right\}$$

$$= \prod_i \left(1 + e^{-\beta(\epsilon_i-\mu)}\right), \tag{5.20}$$

where one has to use the fact that $[\hat{\Psi}_i, \hat{N}_j] = \delta_{ij}\Psi_i$. The expression for Z above is merely the product of Z over the different states i. This means that $\log Z$ will be additive over the different states.

The entropy operator $\hat{S}$ is

$$\hat{S} = -\log\hat{\rho} = \log Z + \beta(\hat{H} - \mu\hat{N})$$

$$= \beta(\hat{H} - \mu\hat{N} - \Omega), \tag{5.21}$$

where the grand canonical potential is defined as $\Omega = -\beta\log Z$. The expectation value of $\hat{S}$ is

$$S = \langle\hat{S}\rangle = \frac{\beta}{Z}\mathrm{Tr}\left\{(\hat{H} - \mu\hat{N})e^{-\beta(\hat{H}-\mu\hat{N})}\right\} + \log Z. \tag{5.22}$$

This may be simplified by noting that

$$\frac{1}{Z} \text{Tr}\left\{ (\hat{H} - \mu \hat{N}) e^{-\beta(\hat{H} - \mu \hat{N})} \right\} = -\frac{\partial \log Z}{\partial \beta}$$

$$= -\frac{\partial}{\partial \beta} \left(\sum_i \log\left(1 + e^{-\beta(\epsilon_i - \mu)}\right) \right)$$

$$= \sum_i \frac{(\epsilon_i - \mu) e^{-\beta(\epsilon_i - \mu)}}{1 + e^{-\beta(\epsilon_i - \mu)}}$$

$$= \sum_i (\epsilon_i - \mu) f(\epsilon_i),$$

$$(5.23)$$

which gives an analogous expression to Eq. (5.13) but with a sum over states

$$S = \sum_i \left(\beta(\epsilon_i - \mu) f_i + \log\left(1 + e^{-\beta(\epsilon_i - \mu)}\right) \right)$$

$$= \sum_i -\left(f_i \log f_i + (1 - f_i) \log(1 - f_i) \right).$$

$$(5.24)$$

The above equation tells us that the entropy is additive for independent fermions which is what we would expect [3].

The above equation for entropy motivates us to write down the density matrix in a form given by Eq. (5.15):

$$\hat{\rho} = \otimes_k \hat{\rho}_k = \otimes_k \begin{pmatrix} 1 - f_k & 0 \\ 0 & f_k \end{pmatrix}. \qquad (5.25)$$

The above density operator represents N uncorrelated systems where N corresponds to the number of energy levels available to a single particle. Each energy level, along with the corresponding vacuum, behaves as an independent system. Each of these Hilbert spaces has two state vectors $\{|0\rangle_n, |1\rangle_n\}$ corresponding to the vacuum and the one-particle level of energy ϵ_n. The entire Hilbert space is then a direct product of such Hilbert spaces and its density operator will have the form given in Eq. (5.25) when these systems are uncorrelated.

In the second quantized notation, we write

$$\hat{\rho} = \prod_k \left((1 - f_k) \hat{\Psi}_k \hat{\Psi}_k^\dagger + f_k \hat{\Psi}_k^\dagger \hat{\Psi}_k \right). \qquad (5.26)$$

The joint entropy theorem [4] tells us that the entropy of uncorrelated systems has to be additive

$$S(\hat{\rho}) = \sum_k S(\hat{\rho}_k) = \sum_k -\left(f_k \log f_k + (1 - f_k) \log(1 - f_k) \right). \qquad (5.27)$$

5.2   Entropy in the Scattering Basis

We now discuss the formulation of entropy in the context of a scattering problem. The system is composed of M reservoirs which exchange particles and energy with each other through a central scattering region. Naturally, this is a nonequilibrium situation where there is a net flux of particles and energy entering and leaving the system and thereby moving it away from equilibrium. However, the reservoirs are taken to be in the thermodynamic limit where it is assumed that their equilibrium configuration does not change. In this limit, the reservoirs set up the boundary conditions for the scattering problem.

We label the reservoirs $\alpha = \{1, 2, \ldots, M\}$ and we consider the scattering problem in the absence of two-body interactions. The total Hamiltonian is given by

$$\hat{H} = \hat{H}_{\text{sys}} + \hat{H}_{\text{res}} + \hat{H}_{\text{s-r}}, \tag{5.28}$$

where [2]

$$\hat{H}_{\text{sys}} = \sum_{i,j} \left(H_{\text{sys}}\right)_{ij} \hat{d}_i^\dagger \hat{d}_j \tag{5.29}$$

is a generic one-body Hamiltonian for a finite spatial domain, with $\left(H_{\text{sys}}\right)_{ij}^* = \left(H_{\text{sys}}\right)_{ji}$,

$$\hat{H}_{\text{res}} = \sum_{\alpha=1}^{M} \sum_{k \in \alpha} \epsilon_k \hat{c}_k^\dagger \hat{c}_k \tag{5.30}$$

is the Hamiltonian describing M fermion reservoirs, and

$$\hat{H}_{\text{s-r}} = \sum_{i} \sum_{\alpha=1}^{M} \sum_{k \in \alpha} \left(V_{ik} \hat{d}_i^\dagger \hat{c}_k + \text{H.c.}\right). \tag{5.31}$$

The most natural framework to calculate quantities in the scattering problem are the scattering basis states [5–7] since it exactly diagonalizes the Hamiltonian in Eq. (5.28). The scattering basis states associated with a reservoir are an orthogonal set of basis states. Furthermore, they are also orthogonal with respect to any other reservoir. We label the reservoirs (corresponding states) as α (state k) and β (state k'), and write the anti-commutation relations

$$\{\hat{\Psi}_{\alpha,k}, \hat{\Psi}_{\beta,k'}^\dagger\} = \delta_{\alpha\beta}\delta_{kk'}. \tag{5.32}$$

---

[2]For clarity of notation, Fock-space operators are written with a hat, while matrices defined in the one-body Hilbert space of the system are written without a hat.

In the scattering basis, the total Hamiltonian Eq. (5.28) and number operators are given by the direct sum

$$\hat{H} = \oplus_\alpha \hat{H}_\alpha$$
$$\hat{N} = \oplus_\alpha \hat{N}_\alpha,$$

(5.33)

where

$$\hat{H}_\alpha = \sum_{k_\alpha} \epsilon_{k_\alpha} \hat{\Psi}^\dagger_{\alpha,k} \hat{\Psi}_{\alpha,k}$$

$$\hat{N}_\alpha = \sum_{k_\alpha} \hat{\Psi}^\dagger_{\alpha,k} \hat{\Psi}_{\alpha,k}$$

(5.34)

are the Hamiltonian and number operators, respectively, associated with reservoir α. They obey the commutation relations

$$[\hat{H}_\alpha, \hat{H}_\beta] = [\hat{N}_\alpha, \hat{N}_\beta] = 0.$$

(5.35)

The density operator in the scattering basis is simply an extension of the density operator given by Eq. (5.26) with an additional product over scattering states corresponding to different reservoirs

$$\prod_k \rightarrow \prod_\alpha \prod_{k_\alpha}.$$

(5.36)

Explicitly,

$$\hat{\rho} = \prod_\alpha \prod_{k_\alpha} \left((1 - f_{k_\alpha}) \hat{\Psi}_{k_\alpha} \hat{\Psi}^\dagger_{k_\alpha} + f_{k_\alpha} \hat{\Psi}^\dagger_{k_\alpha} \hat{\Psi}_{k_\alpha} \right).$$

(5.37)

The density matrices are multiplicative and thereby entropies additive due to the commutation of the different subsystems [cf. Eq. (5.35)] and we directly use Eq. (5.21) to write the entropy operator as

$$\hat{S} = \sum_\alpha \hat{S}_\alpha = \sum_\alpha \beta_\alpha (\hat{H}_\alpha - \mu_\alpha \hat{N}_\alpha - \Omega_\alpha).$$

(5.38)

In the same operator representation, we may even write the differential form of the first law as

$$\delta \hat{S} = \sum_\alpha \delta \hat{S}_\alpha = \sum_\alpha \beta_\alpha (\delta \hat{H}_\alpha - \mu_\alpha \delta \hat{N}_\alpha).$$

(5.39)

The scattering states form a continuum and the entropy operator $\hat{S}_\alpha$ may be written as an integral over energy by introducing the (partial) density of states [6] associated with reservoir α $g_{k_\alpha}(\omega) = \sum_{k_\alpha} \delta(\omega - \epsilon_{k_\alpha})$:

$$
\begin{aligned}
\hat{S}_\alpha &= \sum_k \left((\epsilon_k - \mu_\alpha) \hat{\Psi}^\dagger_{\alpha,k} \hat{\Psi}_{\alpha,k} + \log\left(1 + e^{-\beta_\alpha(\epsilon_k - \mu_\alpha)}\right) \right) \\
&= \int_{-\infty}^{\infty} d\omega g_\alpha(\omega) \left((\omega - \mu_\alpha) \hat{\Psi}^\dagger_\alpha(\omega) \hat{\Psi}_\alpha(\omega) + \log\left(1 + e^{-\beta_\alpha(\omega - \mu_\alpha)}\right) \right).
\end{aligned}
$$
(5.40)

The corresponding entropy then is the expectation value of the above operator

$$
S_\alpha = - \sum_{k_\alpha} \left(f_{k_\alpha} \log f_{k_\alpha} + (1 - f_{k_\alpha}) \log(1 - f_{k_\alpha}) \right),
$$
(5.41)

and it is easy to see that

$$
S_\alpha = - \int_{-\infty}^{\infty} d\omega \, g_\alpha(\omega) \left(f_\alpha(\omega) \log f_\alpha(\omega) + (1 - f_\alpha(\omega)) \log\left(1 - f_\alpha(\omega)\right) \right),
$$
(5.42)

where $f_\alpha \equiv f_\alpha(\omega)$ is the Fermi–Dirac distribution associated with reservoir α. The total entropy is simply the sum of the different reservoir contributions

$$
S = \sum_\alpha S_\alpha.
$$
(5.43)

That is, in the absence of many-body interactions, the scattering states corresponding to different reservoirs do not mix [7] and behave as uncorrelated systems.

Particle Number and Energy

The total energy and particle number of the system are given by their respective expectation values of the operators given in Eq. (5.33). It is easy to see that the energy and particle numbers associated with reservoir α, i.e., the expectation values of the operators given in Eq. (5.34) are

$$
\begin{aligned}
E_\alpha &\equiv \langle \hat{H}_\alpha \rangle = \int_{-\infty}^{\infty} d\omega \, \omega g_\alpha(\omega) f_\alpha(\omega) \\
N_\alpha &\equiv \langle \hat{N}_\alpha \rangle = \int_{-\infty}^{\infty} d\omega g_\alpha(\omega) f_\alpha(\omega).
\end{aligned}
$$
(5.44)

The total energy and particle number of the system are given by

$$E = \sum_\alpha \int_{-\infty}^{\infty} d\omega \, \omega g_\alpha(\omega) f_\alpha(\omega)$$

$$N = \sum_\alpha \int_{-\infty}^{\infty} d\omega g_\alpha(\omega) f_\alpha(\omega). \tag{5.45}$$

Relation to Green's Functions

The (partial) density of states g_α associated with reservoir α can be expressed in terms of the Green's function of the system

$$g_\alpha(\omega) = \frac{1}{2\pi} \mathrm{Tr}\{G^r(\omega)\Gamma^\alpha(\omega)G^a(\omega)\}, \tag{5.46}$$

where Γ^α is the tunneling-width matrix of the system with reservoir α. Note that $\mathrm{Tr}\{.\}$ here refers to the trace over the one-body Hilbert space of the system. The Green's functions are in fact correlation functions which are calculated in the many-body Hilbert space. In reading the above equation, the Green's functions are understood to be matrices (they are not operators in Fock space) acting on the one-body Hilbert space of the system. We refer the reader to the book by Stefanucci and van Leeuwen [8] for details pertaining to the Green's function formalism and Appendix A for relevant details pertaining to the nonequilibrium steady state. The matrix elements of the retarded Green's function are defined by

$$G^r_{ij}(t) = -i\theta(t)\langle\{\hat{d}_i(t), \hat{d}^\dagger_j(0)\}\rangle, \tag{5.47}$$

where the curly brackets $\{.,.\}$ denote the anti-commutation and $\langle.\rangle$ denotes the quantum and statistical average.

Equation (5.46) gives us the partial density of states [6, 9] of the system associated with reservoir α. The total density of states is given by

$$g(\omega) = \sum_\alpha g_\alpha(\omega) = \sum_\alpha \frac{1}{2\pi} \mathrm{Tr}\{G^r(\omega)\Gamma^\alpha(\omega)G^a(\omega)\}$$

$$= \sum_\alpha \mathrm{Tr}\{A_\alpha(\omega)\} \tag{5.48}$$

$$= \mathrm{Tr}\{A(\omega)\},$$

where $A_\alpha(\omega)$ is the partial spectral function associated with reservoir α whereas $A(\omega)$ is the total spectral function.

5.2.1   Entropy Matrix of the System

The Green's functions and the spectral functions are matrices[3] in the one-body Hilbert space of the system $\mathcal{H}_1$ which take into account exactly the open nature of the system via the tunneling-width matrices Γ^α (which are also matrices in $\mathcal{H}_1$) describing the coupling of the system to the reservoir α.

We may then write the entropy matrix $\mathbf{S}$ as

$$\mathbf{S} = -\int_{-\infty}^{\infty} d\omega \sum_\alpha A_\alpha \big(f_\alpha \log f_\alpha + (1 - f_\alpha) \log(1 - f_\alpha) \big) \qquad (5.49)$$

where A_α is the matrix of the partial spectral function introduced in Eq. (5.48). Explicitly:

$$A_\alpha(\omega) = \frac{1}{2\pi} G^r(\omega) \Gamma^\alpha(\omega) G^a(\omega). \qquad (5.50)$$

We suppressed the energy dependence of A_α and f_α in Eq. (5.49) for notational simplicity. The total spectral function was introduced in Eq. (5.48) and, for clarity, we write it explicitly:

$$\begin{aligned} A(\omega) = \sum_\alpha A_\alpha(\omega) &= \frac{1}{2\pi} \sum_\alpha G^r(\omega) \Gamma^\alpha(\omega) G^a(\omega) \\ &= \frac{1}{2\pi} G^r(\omega) \Gamma(\omega) G^a(\omega), \end{aligned} \qquad (5.51)$$

where $\Gamma(\omega) = \sum_\alpha \Gamma^\alpha(\omega)$ is tunneling-width matrix describing the total coupling of the system to the reservoirs. The matrices A_α and Γ^α can be shown to be positive-semidefinite [2, 9, 10]. This implies that the entropy matrix $\mathbf{S}$ is also positive-semidefinite. $A(\omega) = \sum_\alpha A_\alpha(\omega)$ integrates to

$$\int_{-\infty}^{\infty} d\omega \, A(\omega) = \mathbb{1}, \qquad (5.52)$$

the identity matrix in the one-body Hilbert space $\mathcal{H}_1$. The lesser Green's function provides information regarding the nonequilibrium occupancy of the system and obeys the Keldysh equation [8] (also see Appendix B)

$$G^<(\omega) = G^r(\omega) \Sigma^<(\omega) G^a(\omega), \qquad (5.53)$$

---

[3] As mentioned previously, we use the term matrices to highlight the fact that they are not operators in the Fock space but are defined on the one-body Hilbert space of the system.

where $\Sigma^<$ is the "lesser" self-energy. In the absence of two-body interactions

$$\Sigma^<(\omega) = i \sum_\alpha \Gamma^\alpha(\omega) f_\alpha(\omega). \tag{5.54}$$

The lesser Green's function can then be written as

$$G^<(\omega) - 2\pi i \sum_\alpha A_\alpha(\omega) f_\alpha(\omega). \tag{5.55}$$

The lesser and greater have been defined in Appendix A and their relation to the retarded and advanced Green's function is also discussed.

We may similarly define the particle number (**N**) and energy (**E**) matrices in the one-body Hilbert space of the system.

$$\begin{aligned}
\mathbf{N} &= \sum_\alpha \int_{-\infty}^\infty d\omega\, A_\alpha(\omega) f_\alpha(\omega) \\
\mathbf{E} &= \sum_\alpha \int_{-\infty}^\infty d\omega\, \omega A_\alpha(\omega) f_\alpha(\omega).
\end{aligned} \tag{5.56}$$

5.2.2 Entropy of Subspaces

The entropy matrix **S** given by Eq. (5.49) can be projected onto any subspace $\mathcal{A}$ of the one-body Hilbert space $\mathcal{H}_1$. We denote the entropy associated with the subspace $\mathcal{A}$ as $S[\mathcal{A}]$

$$S[\mathcal{A}] \equiv \mathrm{Tr}\{P_\mathcal{A} \mathbf{S}\}, \tag{5.57}$$

where $P_\mathcal{A}$ is the projection operator for the subspace $\mathcal{A}$. We denote the partial density of states associated with the reservoir α and the subspace $\mathcal{A}$ as

$$g_{\mathcal{A},\alpha} \equiv \mathrm{Tr}\{P_\mathcal{A} A_\alpha\} \tag{5.58}$$

so that

$$S[\mathcal{A}] = \sum_\alpha \int_{-\infty}^\infty d\omega\, g_{\mathcal{A},\alpha}\, \sigma(f_\alpha), \tag{5.59}$$

where we introduce the function

$$\sigma(f) = -\big(f \log f + (1 - f) \log(1 - f)\big). \tag{5.60}$$

The energy dependence of $g_{\mathcal{A},\alpha}$ and f_α above have been suppressed for notational simplicity.

The particle number and energy associated with the subspace $\mathcal{A}$ are

$$N[\mathcal{A}] \equiv \text{Tr}\{P_{\mathcal{A}}\mathbf{N}\} = \sum_\alpha \int_{-\infty}^\infty d\omega \, g_{\mathcal{A},\alpha} \, f_\alpha$$

$$E[\mathcal{A}] \equiv \text{Tr}\{P_{\mathcal{A}}\mathbf{E}\} = \sum_\alpha \int_{-\infty}^\infty d\omega \, \omega g_{\mathcal{A},\alpha} \, f_\alpha.$$

$$(5.61)$$

Additivity

The subspace entropy has the important property that it is additive over orthogonal subspaces. The entropy associated with the subspace $\mathcal{C} = \mathcal{A} \oplus \mathcal{B}$ is clearly seen to be

$$S[\mathcal{A} \oplus \mathcal{B}] = S[\mathcal{A}] + S[\mathcal{B}], \tag{5.62}$$

since $P_{\mathcal{C}} = P_{\mathcal{A}} + P_{\mathcal{B}}$ when $P_{\mathcal{A}}P_{\mathcal{B}} = 0$ for the orthogonal subspaces $\mathcal{A}, \mathcal{B}$. The total entropy then can be easily written as a sum of entropies over orthogonal subspaces of the one-body Hilbert space $\mathcal{H}_1$.

We may therefore introduce an entropy density $S(\mathbf{x})$ using the projection operator in the position basis $P(\mathbf{x}) = |\mathbf{x}\rangle\langle\mathbf{x}|$

$$S(\mathbf{x}) = \int_{-\infty}^\infty d\omega \sum_\alpha g_\alpha(\mathbf{x}) \, \sigma(f_\alpha), \tag{5.63}$$

where

$$g_\alpha(\mathbf{x}) = \text{Tr}\{P(\mathbf{x})A_\alpha\} = \langle\mathbf{x}| A_\alpha |\mathbf{x}\rangle \tag{5.64}$$

and since the projection operator $P(\mathbf{x})$ obey the completeness relation

$$\int d^3\mathbf{x} \, |\mathbf{x}\rangle\langle\mathbf{x}| = \mathbb{1}, \tag{5.65}$$

we find that the total entropy of the system is

$$S = \int d^3\mathbf{x} \, S(\mathbf{x}). \tag{5.66}$$

The total entropy above is completely equivalent to the one given by Eq. (5.43).

Relation to the Entanglement Entropy The entropy of a subsystem and the local entropy defined here different from the von Neumann entropy computed from the reduced density matrix of a subsystem. The local entropies defined here are clearly additive and we may simply add them up over the entire Hilbert space to obtain the exact total entropy of the system as shown in Eq. (5.66). By contrast, the von Neumann entropy computed from the reduced density matrix of a subsystem leads to entanglement entropy which is not additive over subsystems

It can be easily seen that the particle number and energy of orthogonal subspaces are additive:

$$N[\mathcal{A} \oplus \mathcal{B}] = N[\mathcal{A}] + S[\mathcal{B}]$$
$$E[\mathcal{A} \oplus \mathcal{B}] = E[\mathcal{A}] + S[\mathcal{B}],$$

(5.67)

and that their sum over the entire one-body Hilbert space (trace of $\mathbf{N}$ and $\mathbf{E}$) gives exactly the total particle number and energy given in Eq. (5.45). We similarly define the particle number ($N(\mathbf{x})$) and energy densities ($E(\mathbf{x})$) using the projection operator in the position basis $P(\mathbf{x}) = |\mathbf{x}\rangle\langle\mathbf{x}|$ and find that the total particle number and energy of the system [see Eq. (5.45)] are given by

$$N = \int d^3\mathbf{x}\, N(\mathbf{x})$$
$$E = \int d^3\mathbf{x}\, E(\mathbf{x}).$$

(5.68)

We note that while the additivity of the particle number density is expected for arbitrary interactions, the energy density is additive only for independent fermions.

5.3 Entropy Inferred from Local Measurements

Local measurements do not have access to information regarding the scattering states of the system. Consider, for example, a one dimensional transport problem where a quantum wire is connected to two reservoirs: L (on the left) and R (on the right). A local observer cannot distinguish between states that are moving left (these are scattering states associated with R) and ones that are moving right (scattering states associated with L). This lack of information leads to an *apparent* mixing of the states corresponding to L and R. We show that this lack of information implies an increase in the observed entropy.

The local observer sees a nonequilibrium distribution which is related to the local projection of $G^<$ on the subspace $\mathcal{A}$ of the one-body Hilbert space $\mathcal{H}_1$ which she has access to. She also observes a local spectrum which she has access to:

$$g_A(\omega) = \mathrm{Tr}\{P_A A(\omega)\}$$

$$= \sum_\alpha \mathrm{Tr}\{P_A A_\alpha(\omega)\} \tag{5.69}$$

$$\equiv \sum_\alpha g_{A,\alpha}.$$

The nonequilibrium distribution function sampled by the local observer in the subspace A is[4]

$$f_A \equiv \frac{\mathrm{Tr}\{P_A G^<(\omega)\}}{2\pi i g_A(\omega)}. \tag{5.70}$$

Using Eq. (5.55) for $G^<$, we find that

$$f_A = \frac{\sum_\alpha g_{A,\alpha} f_\alpha}{g_A}. \tag{5.71}$$

The quantities g_A and f_A above are clearly measurable [11] by scanning probe techniques or near-field photoemission. Experiments generally provide position-local measurements where the projection operator can be written as $P(\mathbf{x}) = |\mathbf{x}\rangle\langle\mathbf{x}|$ but covering a finite region of space

$$P_A \equiv \int_{\mathbf{x}\in A} d^3\mathbf{x} P(\mathbf{x}). \tag{5.72}$$

The nonequilibrium distribution function has been discussed in Refs. [9, 10, 12] and has also been covered in Chap. 2 of the present thesis.

The observer sees the system to be locally out of equilibrium and would believe the system to have the density matrix given by

$$\hat\rho = \otimes_n \hat\rho_n = \otimes_n \begin{pmatrix} 1 - f_A(\epsilon_n) & 0 \\ 0 & f_A(\epsilon_n) \end{pmatrix}. \tag{5.73}$$

The index n runs over the local spectrum of states seen by the observer $g_A(\omega) = \sum_n \delta(\omega - \epsilon_n)$. Therefore, we would expect that the observer would infer an entropy for the subsystem which is different from $S[A]$. The entropy can then be written as

$$S_s[A] = -\int_{-\infty}^{\infty} d\omega\, g_A(\omega)\big(f_A(\omega) \ln f_A(\omega) + (1 - f_A(\omega)) \ln (1 - f_A(\omega))\big). \tag{5.74}$$

---

[4] f_A is a generalization to any subspace A of the nonequilibrium distribution function $f_s(\omega)$, as introduced in Chap. 2, which selects the subspace defined by the probe-system coupling Γ^p.

We write the above entropy with the subscript "s" to differentiate it from the entropy calculated with full knowledge of the scattering basis. Our prior work utilized the same subscript for the local nonequilibrium distribution function f_s and we therefore utilize the same subscript here for notational consistency. The reader may refer to the subscript "s" as "system locally." Succinctly:

$$S_s[\mathcal{A}] = \int_{-\infty}^{\infty} d\omega g_{\mathcal{A}}(\omega)\sigma(f_{\mathcal{A}}). \tag{5.75}$$

If one only had knowledge of the local density of states $g_{\mathcal{A}}$ and the local nonequilibrium distribution function $f_{\mathcal{A}}$, the observer would infer the local particle number and energy to be

$$
\begin{aligned}
N_s[\mathcal{A}] &\equiv \int_{-\infty}^{\infty} d\omega g_{\mathcal{A}}(\omega) f_{\mathcal{A}}(\omega) \\
E_s[\mathcal{A}] &\equiv \int_{-\infty}^{\infty} d\omega\, \omega g_{\mathcal{A}}(\omega) f_{\mathcal{A}}(\omega).
\end{aligned}
\tag{5.76}
$$

It is easy to show that the particle number and energy written above are indeed the same as Eq. (5.61). Therefore, the observer indeed infers the correct particle numbers and energy but clearly a different entropy. Henceforth, we use the notation $N_{\mathcal{A}}$ and $E_{\mathcal{A}}$ for the particle number and energy of subspace $\mathcal{A}$

$$
\begin{aligned}
N_{\mathcal{A}} &\equiv N[\mathcal{A}] = N_s[\mathcal{A}] \\
E_{\mathcal{A}} &\equiv E[\mathcal{A}] = E_s[\mathcal{A}].
\end{aligned}
\tag{5.77}
$$

We note that choosing a completely local measurement with $P_{\mathcal{A}} = P(\mathbf{x})$ gives us the entropy density $S_s(\mathbf{x}) = S_s[\mathcal{A}]$, the particle density $N(\mathbf{x}) = N_{\mathcal{A}}$, and the energy density $E(\mathbf{x}) = E_{\mathcal{A}}$. We then designate the local density of states by $g(\omega; \mathbf{x}) = g_{\mathcal{A}}(\omega)$ and the local nonequilibrium distribution as $f(\omega; \mathbf{x}) = f_{\mathcal{A}}(\omega)$.

5.4 Entropy Inferred from a Probe Measurement

In Chap. 2 we covered in great detail the notion of a probe measurement. We noted that the probe measurement (of temperature and voltage) is simply the following problem: Finding an equilibrium Fermi–Dirac distribution $f_p(\omega)$ which has the same energy and occupancy of the nonequilibrium system it probes.

If the observer only had knowledge of the local density of states $g_{\mathcal{A}}$ and was given a distribution function f_p, the inferred particle number and energy would be

$$N_p[\mathcal{A}] \equiv \int_{-\infty}^{\infty} d\omega g_{\mathcal{A}}(\omega) f_p(\omega)$$

$$E_p[\mathcal{A}] \equiv \int_{-\infty}^{\infty} d\omega \, \omega g_{\mathcal{A}}(\omega) f_p(\omega). \tag{5.78}$$

The probe measurement simply finds the equilibrium Fermi–Dirac distribution $f(\mu_p, T_p; \omega) = f_p(\omega)$ such that

$$N_p[\mathcal{A}] = N_{\mathcal{A}}$$

$$E_p[\mathcal{A}] = E_{\mathcal{A}}, \tag{5.79}$$

where the particle number $N_{\mathcal{A}}$ and energy $E_{\mathcal{A}}$ of the subsystem [see Eq. (5.77)]. Equation (5.79) are the constraints that define the probe temperature and chemical potential which was discussed in Chap. 2 in its most general form. We showed that Eq. (5.79) uniquely determines $f_p(\omega) \equiv f(\mu_p, T_p; \omega)$ *if and only if* both the constraints are imposed. We interpreted this as a consequence of the second law of thermodynamics. The density matrix for the probe distribution may be constructed in exactly the same way as was done in Eq. (5.73):

$$\hat{\rho}_p = \otimes_n \hat{\rho}_n = \otimes_n \begin{pmatrix} 1 - f_p(\epsilon_n) & 0 \\ 0 & f_p(\epsilon_n) \end{pmatrix}, \tag{5.80}$$

where the spectrum $g_{\mathcal{A}}(\omega) = \sum_n \delta(\omega - \epsilon_n)$ with energies ϵ_n in the subspace $\mathcal{A}$ are now filled with an equilibrium distribution function f_p. We may then write the entropy associated with the equilibrium distribution function f_p in the same way as Eq. (5.75):

$$S_p[\mathcal{A}] = -\int_{-\infty}^{\infty} d\omega g_{\mathcal{A}}(\omega)\big(f_p(\omega) \ln f_p(\omega) + \big(1 - f_p(\omega)\big) \ln \big(1 - f_p(\omega)\big)\big)$$

$$= \int_{-\infty}^{\infty} d\omega g_{\mathcal{A}}(\omega)\sigma(f_p). \tag{5.81}$$

Once again, if we talk about purely local measurements $P_{\mathcal{A}} = P(\mathbf{x})$, we shall designate the entropy density measured by the probe as $S_p(\mathbf{x})$. We use the notation $T(\mathbf{x})$ and $\mu(\mathbf{x})$ to designate the local temperature and chemical potential, respectively. We would of course expect the equilibrium entropy S_p to be larger than the corresponding nonequilibrium S_s subject to the same constraints Eq. (5.79). We now discuss the inequalities satisfied by the entropies S, S_s, and S_p.

5.5   Entropy Inequalities

We note that the function σ is concave and thereby obeys the Jensen's inequality [13]:

$$\sigma(\lambda f_1 + (1 - \lambda) f_2) \geq \lambda \sigma(f_1) + (1 - \lambda)\sigma(f_2), \qquad (5.82)$$

as illustrated in Fig. 5.1 When $\lambda \in [0, 1]$, $f = \lambda f_1 + (1 - \lambda) f_2$ is a point on the x-axis between f_1 and f_2 whose function value $\sigma(f)$ is greater than the corresponding point on the line segment shown in red $\lambda \sigma(f_1) + (1 - \lambda)\sigma(f_2)$ which connects points $P_1 = (f_1, \sigma(f_1))$ and $P_2 = (f_2, \sigma(f_2))$. In other words, the blue curve is always above the red line between points P_1 and P_2. The equality holds when $\lambda = 0$ or 1, or when $f_1 = f_2$.

Theorem 5.1 $S_s[\mathcal{A}] \geq S[\mathcal{A}]$.

Proof The proof is a simple application of Jensen's inequality. We may generalize the inequality (5.82) for M points

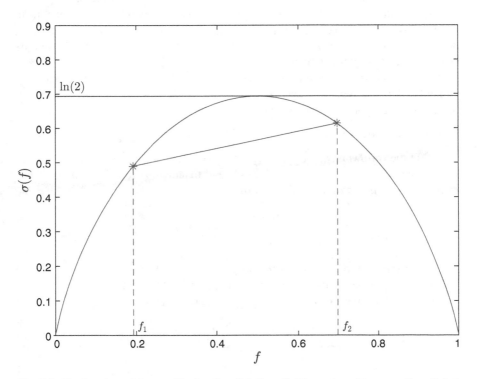

Fig. 5.1 The function $\sigma(f) = -f \ln f - (1 - f) \ln(1 - f)$. Illustration of its concavity: $\sigma(\lambda f_1 + (1 - \lambda) f_2) \geq \lambda \sigma(f_1) + (1 - \lambda)\sigma(f_2)$

$$\sigma\left(\sum_i \lambda_i f_i\right) \geq \sum_i \lambda_i \sigma(f_i), \tag{5.83}$$

where $\sum_i \lambda_i = 1$ and $\lambda_i \geq 0 \; \forall \, i \in \{1, 2, 3 \ldots M\}$. We apply this directly to entropy $S_s[\mathcal{A}]$ in Eq. (5.75)

$$
\begin{aligned}
S_s[\mathcal{A}] &= \int_{-\infty}^{\infty} d\omega \; g_A \sigma(f_A) \\
&= \int_{-\infty}^{\infty} d\omega \; g_A \sigma\left(\frac{\sum_\alpha g_{A,\alpha} f_\alpha}{g_A}\right) \\
&\geq \int_{-\infty}^{\infty} d\omega \sum_\alpha g_{A,\alpha} \sigma(f_\alpha) \equiv S[\mathcal{A}],
\end{aligned}
\tag{5.84}
$$

where we used the fact that $g_A = \sum_\alpha g_{A,\alpha}$ from Eq. (5.69) and also the fact that $g_A f_A = \sum_\alpha g_{A,\alpha} f_\alpha$ as shown in Eq. (5.71). ■

We now consider the entropy by supposing that we do not have access to the nonequilibrium distribution function f_A but do have a method to measure the mean particle number N_A and the mean energy E_A associated with the subsystem $\mathcal{A}$. As stated in the previous section, the probe measurement seeks to find the unique Fermi–Dirac distribution which satisfies this condition Eq. (5.79).

It is well known that the equilibrium distribution $f = f_p$ extremizes the entropy subject to the constraints of fixed particle number and energy. We refer to this result as the maximum entropy principle and show it explicitly below.

Theorem 5.2 (Maximum Entropy Principle) $S_p[\mathcal{A}] \geq S_s[\mathcal{A}]$.

Proof In the following, since we work in the subspace $\mathcal{A}$, we do not specify it for the entropy. We use the square brackets instead to show the functional dependence of the entropy $S[f]$ on the distribution function f. The entropy

$$S[f] = \int_{-\infty}^{\infty} d\omega \; g_A(\omega) \sigma(f) \tag{5.85}$$

is then a functional of f and we would get $S_s[\mathcal{A}] = S[f_A]$ when $f = f_A$ and $S_p[\mathcal{A}] = S[f_p]$ when f is set to the probe distribution function $f = f_p$.

We write the variations in the entropy subject to the constraints of Eq. (5.79) as

$$
\begin{aligned}
\delta S[f] = \delta\Bigg(& \int_{-\infty}^{\infty} d\omega \; g_A(\omega) \sigma(f) \\
& + \lambda_1 \left[\int_{-\infty}^{\infty} d\omega \; g_A(\omega) f(\omega) - N_A\right] + \lambda_2 \left[\int_{-\infty}^{\infty} d\omega \; \omega g_A(\omega) f(\omega) - E_A\right] \Bigg),
\end{aligned}
\tag{5.86}
$$

where λ_1, λ_2 are the Lagrange multipliers associated with the constraints of fixed particle number N_A and fixed energy E_A, respectively. N_A and E_A were defined in Eq. (5.77) and the constraint was explicitly stated in the Sect. 5.4. We find that

$$\delta S[f] = \int_{-\infty}^{\infty} d\omega \, g_A(\omega) \left[-\log\left(\frac{f}{1-f}\right) + \lambda_1 + \omega\lambda_2 \right] \delta f \tag{5.87}$$

and extremizing it ($\delta S = 0$) for any variation δf would imply that the integrand above vanishes and we obtain

$$f = \frac{1}{1 + e^{-\lambda_1 - \omega\lambda_2}}. \tag{5.88}$$

The usual identification of these multipliers is

$$\lambda_1 = \frac{\mu_p}{T_p}, \quad \lambda_2 = -\frac{1}{T_p}. \tag{5.89}$$

It was shown in Chap. 2 that the Fermi–Dirac distribution which has the same particle number and energy is a unique one. We see here that the same Fermi–Dirac distribution extremizes the entropy subject to those constraints. Furthermore, we know that the entropy $S[f]$ is concave. We can explicitly write the second order variation of the entropy

$$\delta^2 S[f] = \int_{-\infty}^{\infty} d\omega \, g_A(\omega) \left[-\frac{1}{f} - \frac{1}{1-f} \right] \delta f^2, \tag{5.90}$$

which is always negative since $0 \le f \le 1$. Therefore, f_p in fact maximizes the entropy subject to the constraints of Eq. (5.79). ∎

We therefore have the hierarchy of inequalities

$$\boxed{S[A] \le S_s[A] \le S_p[A].} \tag{5.91}$$

The equality holds only when all the reservoirs are maintained at the same temperature and chemical potential, i.e., in equilibrium. The inequality is illustrated for a two-level system out of equilibrium in Fig. 5.2. We provide some details regarding the two-level system in Sect. 5.5.

Table 5.1 summarizes the relationship between the observed entropy and the information available to the observer.

We therefore find that full knowledge of the scattering states of the system leads to the least entropy $S[A]$ for any subspace A. Indeed, this entropy vanishes when all the reservoirs are maintained at zero temperature, consistent with the third law of thermodynamics. However, the system may still appear to be out

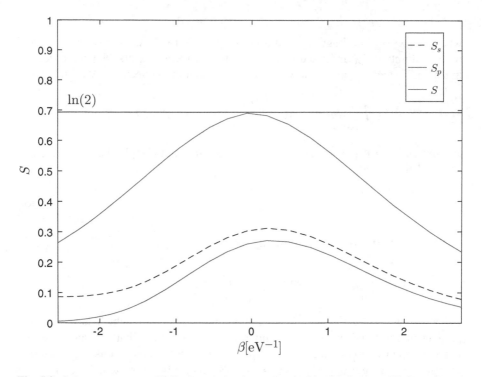

Fig. 5.2 Subsystem entropy $S[\mathcal{A}]$ of a two-level quantum system far from equilibrium, plotted versus the inverse temperature $\beta = (k_B T_p)^{-1}$ of the probe. The electrical bias across the system was varied from 1.6 to 3.2 V. Values of $\beta < 0$ correspond to absolute negative temperatures (population inversion). We see that $S[\mathcal{A}] \leq S_s[\mathcal{A}] \leq S_p[\mathcal{A}]$. Numerical details pertaining to the two-level system have been included at the end of Sect. 5.5

of equilibrium for a local observer. If the observer has access only to the local distribution but no knowledge of the scattering states, it would appear to the observer that the distribution does not correspond to the one at zero temperature. A local observer with access only to the local distribution always sees an increased entropy. Furthermore, if the observer has no idea about the local distribution but only knows the temperature and voltage by performing local probe measurements, the entropy increases even further. We therefore see the intimate relationship between the access to information pertaining to the system and the observed entropy.

Two-Level System: Some Details
We provide here some details regarding the two-level system used to illustrate the hierarchy of inequalities Eq. (5.91) in Fig. 5.2. We label the sites of the two-level system as $|1\rangle$ and $|2\rangle$ which are taken as orthogonal and, in this basis, the Hamiltonian was taken to be

Table 5.1 Table summarizing the notation and meaning of the different entropies based on the information available to the observer

Notation	Meaning	Information Available (from the measurement)
S	Entropy Matrix (in $\mathcal{H}_1$)	Complete Information of the Scattering States
$S[\mathcal{A}]$	Projection of S in subspace $\mathcal{A}$	Complete Information of the Scattering States
$S_s[\mathcal{A}]$	subscript s for "system locally" in subspace $\mathcal{A}$	Local DOS known Local DF known
$S_p[\mathcal{A}]$	subscript p for "probe"	Local DOS known Local DF unknown N_A and E_A known

The abbreviations DOS and DF stand for density of states and distribution function, respectively. The entropies are inversely related to the information that is available to the local observer. The most informative case ($S[\mathcal{A}]$) has been colored in green, the intermediate one in yellow ($S_s[\mathcal{A}]$), and the least informative in red ($S_p[\mathcal{A}]$). Equilibrium probe entropy $S_p[\mathcal{A}]$ is the greatest of the three entropies. In this case, the observer does not know the local distribution function but does know the local density of states and first two moments of the local energy densities (N_A and E_A)

$$H = \begin{pmatrix} \epsilon_0 & V \\ V^* & -\epsilon_0 \end{pmatrix}, \tag{5.92}$$

where $\epsilon_0 = \sqrt{(62)}/8$ and $V = (1 - i)/8$ so that the eigenvalues are $\epsilon_\pm = \pm 1$, and the coupling to the two sites were $\Gamma^{(1)} = \mathrm{diag}(0.15, 0)$ and $\Gamma^{(2)} = \mathrm{diag}(0, 0.15)$. The energy units were taken in [eV]. The equilibrium chemical potential was set to zero. The subspace $\mathcal{A}$ was defined by the local projection taken on the first site:

$$P_\mathcal{A} = |1\rangle\langle 1|. \tag{5.93}$$

The figure employs a symmetric voltage bias, while the temperature of both reservoirs was set to $T = 300\,\mathrm{K}$.

The voltage bias $\delta\mu = \mu_1 - \mu_2$ was varied from 1.6 to 3.2 eV, but note that the figure shows the variation of entropy with respect to the inverse temperature of the local subsystem (and not the bias voltage itself). The temperature (T_p) and voltage (μ_p) of the local subsystem are calculated by solving the nonlinear system of equations given by Eq. (5.79), and we use Newton's root finding method to solve this equation.

5.6   Third Law of Thermodynamics

We discuss here the formulation of the third law of thermodynamics in open quantum systems both in equilibrium and in nonequilibrium steady states. Our formulation of the third law of thermodynamics is in the context of the scattering problem in the absence of two-body interactions, described by the Hamiltonian in Eq. (5.28).

We show [14] that *fully open quantum systems*, defined as those in which all the levels are broadened by the coupling to the environment, obey the strong form of the third law: $S(T) \to 0$ as $T \to 0$, where T is the temperature of the reservoir in thermal equilibrium with the system. Generic open quantum systems are shown to obey $S(T) \to g \ln 2$ as $T \to 0$, where g is the number of localized states lying exactly at the chemical potential of the reservoir. For driven open quantum systems in a nonequilibrium steady state, it is shown that the local entropy $S(\mathbf{x}; T) \to 0$ as $T(\mathbf{x}) \to 0$, except for cases of measure zero arising due to localized states, where $T(\mathbf{x})$ is the temperature measured by a local thermometer as discussed in Sect. 5.4. This latter result holds, of course, for any subspace $\mathcal{A}$ of the one-body Hilbert space but we work in the position-local subspace for compactness of notation.

5.6.1   Background

The third law of thermodynamics has two different formulations, both due to Nernst [15]: (A) The Nernst heat theorem, which states that the equilibrium entropy of a pure substance goes to zero at zero temperature, and (B) The unattainability principle, which states that it is impossible to cool any system to absolute zero in a finite number of operations.

There has been some skepticism about the status of the third law of thermodynamics in the context of open quantum systems in recent literature. We therefore first give some background regarding this discussion. For example, the unattainability principle (B) was challenged in Ref. [16], with claims that zero temperature can be reached but that formulation (A) still holds true. Reference [17] arrives at a result which is in violation of the unattainability principle as pointed out in a comment by Kosloff [18].

Kosloff and collaborators [19] consider the unattainability principle (B) in the context of quantum absorption refrigerators, and show that it is not possible to cool to absolute zero in finite time. The authors warn that the quantum Master equation has to be used carefully, and that violations of the laws of thermodynamics could result otherwise. Kosloff advocates for a careful use of the Master equation and in Ref. [20] argues that such apparent violations [16] of the laws of thermodynamics are caused by uncontrolled approximations. Reference [21] expresses the third law as a no-go theorem for ground-state cooling and similarly encourages caution while using approximations in the quantum Master equation. Reference [22] provides a proof of the unattainability principle (B) using quantum resource theory, and clarifies its connection to the heat theorem (A).

Statement (A) of the third law was proven for a quantum oscillator in contact with various types of heat baths in Refs. [23] and [24]. In Ref. [24], O'Connell rebuts the early claims of the violations of the laws of thermodynamics made in the field of quantum thermodynamics. In particular, he focuses on Ref. [25], which claims to construct a perpetual motion machine. As relates to the third law, Ref. [?5] argues that "neither the von Neumann entropy nor the Boltzmann entropy vanishes when the bath temperature is zero," leading to a claim of a violation of the third law for nonweak coupling. O'Connell calculates the von Neumann entropy [24] and points out that when the interaction energy is considerable, the von Neumann formula can only be applied to the entire system and not to the reduced system.

Perhaps the most flagrant violation of the third law of thermodynamics was put forward by Esposito et al. [26], who claim not only that Nernst heat theorem does not hold, but that the entropy of an open quantum system is *undefined* in the limit of zero temperature.[5] Their approach is inspired by that of Sanchez and coworkers [27, 28], who argue that the definition of heat in open quantum systems is ambiguous when time-dependent driving is present. In Ref. [27], they argue that in non-steady states, the tunneling region has some energy ("energy reactance") and it is unclear whether to ascribe that to the "system" or the "bath." This, they argue, leads to an ambiguity in the definition of the heat, which they propose is fixed by ascribing half the energy of the tunneling region to the bath. Although this definition of heat does agree with the laws of thermodynamics, it is not clear whether their prescription is applicable to models other than the one they consider.

In Ref. [29], Nitzan and collaborators consider a driven resonant level model, and show that the problem of separately defining "system" and "bath" in the strong-coupling regime is circumvented by considering as the system everything that is influenced by the externally driven energy level, and rebut the claims of a violation of the heat theorem put forward in Ref. [26]. Their book-keeping [29] is similar to that originally put forward in the equilibrium case by Friedel [30]. Finally, Ref. [31] uses a scattering approach to similarly circumvent the problem of system/bath definitions for adiabatically driven open quantum systems, expressing changes in the entropy of the system in terms of asymptotic observables at infinity.

Here we consider both the equilibrium case and the case of a nonequilibrium steady state. We consider a general partitioning of the entire system into subsystem and reservoir(s), where the subsystem can be any finite subspace of the total Hilbert space. We show that no ambiguity arises from the partitioning either in equilibrium or in a nonequilibrium steady state, and give proofs of the heat theorem for both cases.

---

[5]Furthermore, in their attempt to extend the notion of heat to the nonequilibrium setting, Esposito et al. [26] seem to be under the misapprehension that heat is a state function in standard thermodynamics.

5.6.2 Localized States

Our discussions thus far in this chapter implicitly assumed fully open quantum systems. The total spectral function was the sum of the partial spectral functions corresponding to different reservoirs α as expressed in Eq. (5.51). The total density of states, similarly, was given by the sum of the partial density of states over the different reservoirs α as expressed in Eq. (5.48).

However, generic open quantum systems may have localized states which are not broadened by the coupling to the reservoirs. In this case, Eq. (5.51) takes the form

$$A(\omega) = \sum_{\alpha=1}^{M} A_\alpha(\omega) + \sum_{\ell} |\ell\rangle\langle\ell|\delta(\omega - \omega_\ell), \qquad (5.94)$$

where A_α, given by Eq. (5.50), is the partial spectral function associated with reservoir α. The second term on the r.h.s. above denotes the spectral contribution due to *localized states*. The sum over ℓ includes any localized states that are not broadened due to the coupling with the reservoir(s).

The density of states expressed in Eq. (5.48) then takes the form

$$g(\omega) = \mathrm{Tr}\{A(\omega)\} = g_{\mathrm{reg}}(\omega) + \sum_{\ell} \delta(\omega - \omega_\ell), \qquad (5.95)$$

where

$$g_{\mathrm{reg}}(\omega) = \sum_{\alpha=1}^{M} g_\alpha(\omega) \qquad (5.96)$$

is the non-singular part of the spectrum, and [6, 9, 32]

$$g_\alpha(\omega) = \mathrm{Tr}\{A_\alpha(\omega)\} \qquad (5.97)$$

is the partial density of states of the system due to scattering states incident on the system from reservoir α.

It should be emphasized that an open quantum system is a subsystem of a larger system, and $g(\omega)$ gives the energy spectrum of the whole system projected onto the Hilbert space of the subsystem. Other prescriptions for partitioning into subsystem and environment are also possible [29, 30].

Similarly, the local[6] density of states is given by

---

[6]Although we use the position-local subspace in this section, the results of course hold for any subspace of the one-body Hilbert space $\mathcal{H}$ of the system. In this section, we also drop the subscript in $\mathcal{H}_1$, for brevity of notation, to mean the one-body Hilbert space of the system.

$$g(\omega; \mathbf{x}) = \langle \mathbf{x} | A(\omega) | \mathbf{x} \rangle$$

$$= \sum_{\alpha=1}^{M} g_\alpha(\omega; \mathbf{x}) + \sum_\ell |\psi_\ell(\mathbf{x})|^2 \delta(\omega - \omega_\ell), \tag{5.98}$$

where the local partial density of states associated with reservoir α is [6, 9, 32]

$$g_\alpha(\omega; \mathbf{x}) = \langle \mathbf{x} | A_\alpha(\omega) | \mathbf{x} \rangle. \tag{5.99}$$

Definition 5.1 A fully open quantum system has no localized states.

Therefore, for a *fully open quantum system*, the total spectral function is given by Eq. (5.51) and the total density of states in Eq. (5.95) is simply $g(\omega) = g_{\text{reg}}(\omega)$.

Entropy

The entropy for a generic open quantum system with localized states takes the form

$$S = \sum_{\alpha=1}^{M} \int d\omega \, g_\alpha(\omega) \sigma(f_\alpha(\omega)) + \sum_\ell \sigma(f_\ell), \tag{5.100}$$

where as usual $f_\alpha(\omega)$ is the Fermi–Dirac distribution of reservoir α and f_ℓ is the occupancy of the ℓth localized state. $\sigma(f) = -f \log(f) + (1 - f) \log(1 - f)$ which was introduced earlier in Eq. (5.60).

The local nonequilibrium entropy density is

$$S(\mathbf{x}) = \sum_{\alpha=1}^{M} \int d\omega \, g_\alpha(\omega; \mathbf{x}) \sigma(f_\alpha(\omega)) + \sum_\ell |\psi_\ell(\mathbf{x})|^2 \sigma(f_\ell), \tag{5.101}$$

which satisfies $S = \int_{\text{sys}} d^3x \, S(\mathbf{x})$, with S the nonequilibrium entropy given by Eq. (5.43).

Nonequilibrium Distribution

The lesser Green's function in Eq. (5.55) has an additional term in the presence of localized states:

$$G^<(\omega) = 2\pi i \left[\sum_{\alpha=1}^{M} A_\alpha(\omega) f_\alpha(\omega) + \sum_\ell f_\ell |\ell\rangle \langle \ell| \delta(\omega - \omega_\ell) \right], \tag{5.102}$$

where $f_\alpha(\omega)$ and f_ℓ are the Fermi–Dirac distribution of reservoir α and the nonequilibrium occupancy of the ℓth localized state, respectively. The local nonequilibrium distribution function of the system then becomes

$$f(\omega; \mathbf{x}) \equiv \frac{\langle \mathbf{x} | G^<(\omega) | \mathbf{x} \rangle}{2\pi i g(\omega; \mathbf{x})} \tag{5.103}$$

$$= \begin{cases} f_\ell, & \omega = \omega_\ell \\ \frac{1}{g(\omega;\mathbf{x})} \sum_\alpha g_\alpha(\omega; \mathbf{x}) f_\alpha(\omega), & \omega \neq \omega_\ell \end{cases}$$

Note that $f(\omega; \mathbf{x})$ may be discontinuous at $\omega = \omega_\ell$.

The mean occupancy and energy of the system locally can be written in the same way as before [cf. Eq. (5.77)]. Since we use the position-local basis, we designate the mean local occupancy[7] and energy by

$$N(\mathbf{x}) = \int d\omega\, g(\omega; \mathbf{x}) f(\omega, \mathbf{x}),$$

$$E(\mathbf{x}) = \int d\omega\, \omega g(\omega; \mathbf{x}) f(\omega, \mathbf{x}). \tag{5.104}$$

We note again that the expression for the particle density holds quite generally but that the expression for the energy density is only valid for independent fermions.

Sufficient Condition for a Fully Open Quantum System

A sufficient condition for a fully open quantum system is that the total tunneling-width matrix [defined in Eq. (5.51)] satisfy

$$\Gamma(\omega)|\psi\rangle \neq 0 \quad \forall\, |\psi\rangle \in \mathcal{H}, \tag{5.105}$$

where $\mathcal{H}$ denotes the Hilbert space of H_{sys}. Equation (5.105) holds if rank$\{\Gamma\} = \dim \mathcal{H}$. Figure 5.3 (right panel) illustrates the case of a fully open quantum system, for the tight-binding model of a benzene ring coupled to a reservoir, but with $\Gamma = \gamma \mathbb{1}$, so that rank$\{\Gamma\} = 6 = \dim \mathcal{H}$. Note the absence of any localized states in the entropy spectrum.

Equation (5.105) is a sufficient condition for a fully open quantum system, but is not a necessary condition [14].

---

[7] The mean local occupancy is in fact the particle density and similarly the mean local energy is the energy density since, in the position-local basis, the projection operator $P(\mathbf{x})$ obeys Eq. (5.65).

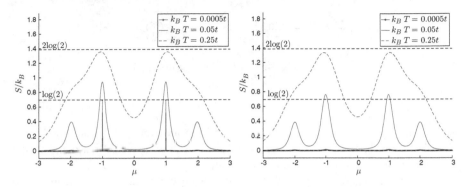

Fig. 5.3 Entropy S of an open quantum system consisting of a benzene ring coupled to an equilibrium electron reservoir, plotted as a function of the chemical potential μ of the reservoir for several temperatures. The system Hamiltonian is taken within the tight-binding model and energies are expressed in units of the nearest-neighbor interaction t. Left panel: Generic open system with $\Gamma_{11} = t$ and all other matrix elements of Γ zero, illustrating the effect of the localized states at $\mu/t = \pm 1$. Right panel: Fully open quantum system with $\Gamma = (t/6)\mathbb{1}$, illustrating the strong form of the third law, $S(T) \to 0$ as $T \to 0 \,\forall \mu$

5.6.3 Statements of the Third Law

For a fermion system coupled to a single reservoir at temperature T and chemical potential μ, we have the following statement of the third law of thermodynamics.

Theorem 5.3 (Third Law of Thermodynamics) *For an open quantum system with a finite-dimensional Hilbert space,*

$$\lim_{T \to 0} S(\mu, T) = 0 \tag{5.106}$$

almost everywhere for $\mu \in \mathcal{R}$.

Proof We consider the first term on the r.h.s. of Eq. (5.100):

$$\lim_{T \to 0} S_{\text{reg}}(\mu, T) = \lim_{T \to 0} \int_{-\infty}^{\infty} d\omega g_{\text{reg}}(\omega) \sigma(f(\omega)), \tag{5.107}$$

where $f(\omega)$ is the Fermi–Dirac distribution of the single reservoir at temperature T and chemical potential μ coupled to the system. We note that

$$\int_{-\infty}^{\infty} d\omega \, g_{\text{reg}}(\omega) = N_{\text{reg}} \leq N_{\mathcal{H}}, \tag{5.108}$$

where $N_{\mathcal{H}} = \dim\{\mathcal{H}\}$ is the dimension of the Hilbert space $\mathcal{H}$ of the system. The Fermi function $\lim_{T \to 0} f(\omega) \to 1 - \Theta(\omega - \mu)$, where Θ is the Heaviside step function. Therefore,

$$\lim_{T \to 0} S_{\text{reg}}(\mu, T) = \lim_{f \to 1} \int_{-\infty}^{\mu} d\omega \, g_{\text{reg}}(\omega) \sigma(f)$$

$$+ \lim_{f \to 0} \int_{\mu}^{\infty} d\omega \, g_{\text{reg}}(\omega) \sigma(f) \tag{5.109}$$

$$= 0,$$

since $\lim_{f \to 1} \sigma(f) = \lim_{f \to 0} \sigma(f) = 0$ and the integral of $g(\omega)$ is bounded by the dimension of the Hilbert space Eq. (5.108). A similar result, restricted to the *resonant level model*, was derived in Ref. [29].

The second term from Eq. (5.100) has the entropy contribution from the localized states which vanish as $T \to 0$ when $\mu \neq \omega_\ell$ since

$$\lim_{f \to 0} \sigma(f(\omega_\ell)) = \lim_{f \to 1} \sigma(f(\omega_\ell)) = 0, \tag{5.110}$$

and when $\mu = \omega_\ell$ we get

$$S_{\text{loc}} = \lim_{T \to 0} \sigma(f(\mu = \omega_\ell)) = \log(2), \tag{5.111}$$

and if there are multiple localized states at $\omega = \omega_\ell$, we denote the degeneracy as g_ℓ and may write

$$S_{\text{loc}} = g_\ell \log(2). \tag{5.112}$$

The points $\mu = \omega_\ell$ have zero measure for $\mu \in \mathcal{R}$ and this completes the proof. ∎

Theorem 5.3 constitutes the general form of the *third law of thermodynamics* for open quantum systems, and is the central result of this section. Figure 5.3 illustrates the behavior of the entropy[8] $S(\mu, T)$ as $T \to 0$ for a model open quantum system consisting of a benzene ring coupled to an electron reservoir.

Corollary 5.3.1 *For a fully open quantum system with a finite-dimensional Hilbert space,*

$$\lim_{T \to 0} S(\mu, T) = 0 \ \forall \mu. \tag{5.113}$$

---

[8]The y-axis in Fig. 5.3 indicates S/k_B, where k_B is the Boltzmann constant. It has been set to unity $k_B = 1$ in all the definitions of the entropies appearing in this chapter including in Sect. 5.7. However, whenever we cite numerical values for the temperatures in [K] (e.g., in Fig. 5.4), it is understood that its conversion to the appropriate energy units ([eV] for our systems) is accompanied by the appropriate numerical value of k_B.

Proof For a fully open system, Eq. (5.108) holds with $N_{\text{reg}} = N_{\mathcal{H}}$ and there are no localized states. Proof follows directly from Eq. (5.109) and Theorem 5.3 holds $\forall \mu \in \mathcal{R}$. ∎

Corollary 5.3.2 *For a finite open quantum system in a nonequilibrium steady state, i.e., when it is coupled to multiple reservoirs at temperatures T_α and chemical potentials μ_α where $\alpha = \{1, 2, \ldots M\}$,*

$$\lim_{T_\alpha \to 0 \,\forall \alpha} S(\{\mu_\alpha\}, \{T_\alpha\}) = 0 \tag{5.114}$$

almost everywhere for $\{\mu_\alpha\} \in \mathcal{R}^M$.

Proof Follows immediately from Theorem 5.3. ∎

We now provide a statement of the third law for the entropy density $S(\mathbf{x})$ in terms of the local temperature $T(\mathbf{x})$ which is defined by an equilibrium Fermi–Dirac distribution having the same local particle and energy densities as the system (see Sect. 5.4).

Theorem 5.4 (Third Law for Nonequilibrium Systems) *For an open quantum system in a nonequilibrium steady state, and with a finite-dimensional Hilbert space,*

$$S(\mathbf{x}) \to 0 \ as \ T(\mathbf{x}) \to 0 \tag{5.115}$$

almost everywhere in space.

Proof $S(\mathbf{x})$ is defined by Eq. (5.101). The contributions of any localized states to $S(\mathbf{x})$ and $S_s(\mathbf{x})$ are identical. Therefore, any difference between $S(\mathbf{x})$ and $S_s(\mathbf{x})$ is due to scattering states. From the concavity of the function $\sigma(f)$ defined by Eq. (5.60), it follows that $S_s(\mathbf{x}) \geq S(\mathbf{x})$, because

$$s(f) \geq \sum_{\alpha=1}^{M} \lambda_\alpha \sigma(f_\alpha), \tag{5.116}$$

where

$$f(\omega; \mathbf{x}) = \sum_{\alpha=1}^{M} \lambda_\alpha(\omega; \mathbf{x}) f_\alpha(\omega), \quad \sum_{\alpha=1}^{M} \lambda_\alpha = 1, \tag{5.117}$$

and $\lambda_\alpha(\omega; \mathbf{x}) = g_\alpha(\omega; \mathbf{x})/g(\omega; \mathbf{x}) \geq 0$. But $S_s(\mathbf{x}) \leq S_p(\mathbf{x})$ by the maximum entropy principle (Theorem 5.2). Therefore, $S(\mathbf{x}) \leq S_p(\mathbf{x})$. ∎

5.7   Entropy as a Metric for Local Equilibrium Departure

We posit that the local entropy difference can be used as a metric quantifying local equilibrium departure. Figure 5.2 shows that the entropy of the auxiliary equilibrium system $S_p[\mathcal{A}]$ follows the local entropy $S_s[\mathcal{A}]$ as well as the entropy of the scattering states $S[\mathcal{A}]$. The probe entropy $S_p[\mathcal{A}]$ reaches a maximum value at infinite temperature which also corresponds, more or less, to the maximum of $S_s[\mathcal{A}]$ as measured by the local observer without knowledge of the scattering states as well as the maximum value of entropy $S[\mathcal{A}]$ evaluated with complete knowledge of the scattering states. Local experiments are capable of measuring $S_p[\mathcal{A}]$ and $S_s[\mathcal{A}]$ of a subspace $\mathcal{A}$. $\mathcal{A}$ is usually a position-local subspace: Scanning tunneling microscopy can be used to measure the local temperature and chemical potential and thereby calculate $S_p[\mathcal{A}]$; photo-emission measurements can be used to measure the local nonequilibrium distribution f_s and thereby calculate the associated entropy $S_s[\mathcal{A}]$. We propose that the entropy difference

$$\Delta S = S_p - S_s \geq 0, \tag{5.118}$$

is a suitable metric to quantify the departure from equilibrium. Clearly, when all reservoirs are maintained at the same temperature and chemical potential, i.e., at equilibrium,

$$\Delta S = 0. \tag{5.119}$$

However, it becomes important to appropriately normalize the entropy for the subspace which we shall discuss in Sect. 5.7.1.

Scanning tunneling measurements can be used to measure the local density of states. In this context, we stated previously [cf. Eq. (3.5)] that the density of states seen by the probe is

$$\bar{A}(\omega) \equiv \frac{\text{Tr}\{\Gamma^p(\omega)A(\omega)\}}{\text{Tr}\{\Gamma^p(\omega)\}},$$

and the local nonequilibrium distribution seen by the probe [see also Eq. (2.3)] is

$$f_s(\omega) \equiv \frac{\text{Tr}\{\Gamma^p(\omega)G^<(\omega)\}}{2\pi i\,\text{Tr}\{\Gamma^p(\omega)A(\omega)\}},$$

which can be measured using photo-emission experiments. It is shown in Appendix A that $0 \leq f_s \leq 1$. We note that the above expressions are completely valid in the presence of arbitrary interactions within the quantum system.

The expressions for the entropy seen by the local observer who has the knowledge of the local nonequilibrium distribution is then given by

$$S_s \equiv S[f_s(\omega)] = -\int_{-\infty}^{\infty} d\omega \bar{A}(\omega)[f_s(\omega) \ln f_s(\omega) + (1 - f_s(\omega)) \ln (1 - f_s(\omega))],$$

$$(5.120)$$

and the entropy inferred from a probe measurement is

$$S_p \equiv S[f_p(\omega)] = -\int_{-\infty}^{\infty} d\omega \bar{A}(\omega)[f_p(\omega) \ln f_p(\omega) + (1 - f_p(\omega)) \ln (1 - f_p(\omega))].$$

$$(5.121)$$

Following the arguments of the previous sections, it can be shown that

$$S_p \leq S_s \qquad\qquad (5.122)$$

even in the presence of interactions. Again, the equality holds only at equilibrium.
    For sufficiently low probe temperatures,

$$S_s \leq S_p \simeq \frac{\pi^2}{3} \bar{A}(\mu_0) k_B T_p, \qquad\qquad (5.123)$$

the leading order term in the Sommerfeld expansion of Eq. (5.121). It is easy to see that the local entropy $S_s \leq S_p \to 0$ as the probe temperature approaches absolute zero $T_p \to 0$, consistent with the *third law of thermodynamics*.
    It is important to note that Eqs. (5.120) and (5.121) are valid for arbitrary interactions within the quantum system. However, their theoretical basis is not at the same foundational level as that of their corresponding entropies in Eqs. (5.75) and (5.81), respectively, which were derived for the noninteracting case. The latter expressions can be compared directly with the exact local entropy of the system in Eq. (5.57) which was derived rigorously in terms of the scattering states. The presence of interactions causes the scattering states to mix and therefore makes their generalization considerably harder [7]. We therefore note that the entropy Eq. (5.120) given in this section is a working ansatz whenever interactions are present.

5.7.1   Per-State Entropy Deficit

We argued that the local entropy deficit $\Delta S = S_p - S_s$ is a suitable metric quantifying the departure from equilibrium. However, it is important to note that the mean local spectrum $\bar{A}(\omega)$ varies significantly from point to point within the nanostructure depending upon the local probe-system coupling (especially in the tunneling regime) and limits the use of ΔS while comparing the "distance" from equilibrium for points within the nanostructure. The situation is analogous to that of a dilute gas, which can have a very low entropy per unit volume even if it has a very high entropy per particle. We note that states far from the equilibrium Fermi energy

μ_p contribute negligibly to the entropy since $\lim_{f \to 0} S[f] = \lim_{f \to 1} S[f] = 0$, and therefore introduce a normalization averaged over the thermal window of the probe:

$$\mathcal{N} = \int_{-\infty}^{\infty} d\omega \frac{\bar{A}(\omega)}{\text{Tr}\{A(\omega)\}} \left(\frac{-\partial f_p}{\partial \omega} \right). \tag{5.124}$$

We define the local entropy-per-state of the system s_s and that of the corresponding local equilibrium distribution s_p as

$$s_s = \frac{S_s}{\mathcal{N}}, \tag{5.125}$$

$$s_p = \frac{S_p}{\mathcal{N}}. \tag{5.126}$$

$\Delta s = s_p - s_s$ quantifies the per-state "distance" from local equilibrium. We present numerical calculations of the local entropy-per-state below and discuss its implications.

5.7.2 Numerical Results

The local temperature distributions shown in Figs. 3.1, 3.2 and 3.3 (presented in Chap. 3) are essentially outside the scope of linear response theory [33] since the cold reservoir $R1$ is held at $T_1 = 0\,\text{K}$, and derivatives of the Fermi function are singular at $T = 0\,\text{K}$. However, it is an open question *how far out of equilibrium* these systems are and which regions therein manifest the most fundamentally nonequilibrium character. To address such questions quantitatively, we use the concept of local entropy-per-state introduced here. In particular, the normalized local entropy deficit $\Delta s \equiv s_p - s_s$ allows us to quantify how far the system is from local equilibrium.

Figure 5.4 shows the local entropy distribution of the system s_s and that of the corresponding local equilibrium distribution s_p, defined by Eqs. (5.125) and (5.126), respectively, for the Au-pyrene-Au junction considered in Chap. 3, Fig. 3.2. The s_p distribution strongly resembles the temperature distribution shown in Fig. 3.2, consistent with the fact [cf. Eq. (5.123)] that the equilibrium entropy of a system of fermions is proportional to temperature at low temperatures. This resemblance is only manifest in the properly normalized entropy-per-state s_p; the spatial variations of S_p are much larger, and stem from the orders-of-magnitude variations of the local density of states $\bar{A}(\mu_0)$. The nonequilibrium entropy distribution s_s of the system qualitatively resembles s_p, but everywhere satisfies the inequality $s_s \leq s_p$. $s_s \to 0$ whenever $T_p \to 0$, consistent with the third law of thermodynamics.

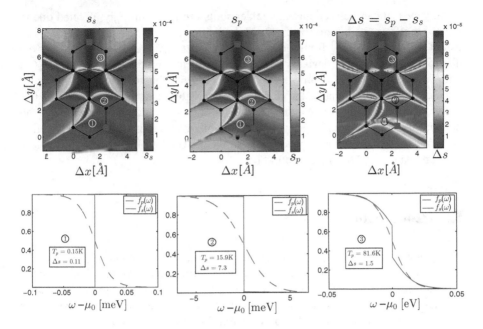

Fig. 5.4 Upper panels (left to right): The local entropy-per-state of the system s_s, of the corresponding local equilibrium distribution s_p, and the local entropy deficit $\Delta s \equiv s_p - s_s$. The temperature distribution for the same junction (with identical bias conditions and sampling of probe positions) is shown in Fig. 3.2, and we note that it resembles almost exactly the distribution s_p. Lower panels: The distributions f_s and f_p for three points shown in the upper panels, each having different probe temperatures $T_p = 0.15$ K, 15.9 K and 81.6 K, respectively. The corresponding entropy deficits are $\Delta s = 0.11$, 7.3, and 1.5, respectively, $\times 10^{-5}$. Point 2, although closer to 0 K than point 3 is to 100 K, is further from local equilibrium

The deviation from local equilibrium quantified by the local entropy deficit $\Delta s = s_p - s_s$ is shown in the top right panel of Fig. 5.4. Δs shows deep blue regions (low entropy deficit) in both the hottest and coldest parts of the system, while the largest entropy deficits (bright red) occur in the areas at intermediate temperatures. This may be explained as follows: within elastic transport theory, the local nonequilibrium distribution function is a linear combination of the distribution functions of the various reservoirs (refer to Appendix C on elastic transport). The entropy deficit is minimal when this distribution function strongly resembles the equilibrium Fermi–Dirac distribution of one of these reservoirs. Conversely, the entropy deficit is maximal when there is a large admixture of both hot and cold electrons without inelastic processes leading to equilibration. Therefore, the hottest and coldest spots show the smallest entropy deficits since there is very little mixing from the cold reservoir $R1$ and hot reservoir $R2$, respectively, while the regions at intermediate temperatures have the largest entropy deficits, and hence are farthest from local equilibrium.

However, it can be seen that the colder spots are more strongly affected due to the mixing from the hot reservoir $R2$, while the hotter spots are affected to a lesser extent due to the mixing from the cold reservoir $R1$. This reflects the fact that the distribution function f_s deviates much more from the distribution function $f_1(\omega)$ with $T_1 \rightarrow 0$ (implying a pure state with zero entropy) due to a small admixture of hot electrons from $R2$ than is the case for the opposite scenario. In other words, it is easier to increase the entropy deficit Δs of a cold spot by adding hot electrons (and thus driving it out of equilibrium) than it is the other way round. The entropy deficit is a good metric to capture such a change in the distribution function and gives us a per-state "distance" from local equilibrium.

5.8   Conclusions

We systematically developed a notion for the nonequilibrium entropy in terms of the scattering states for a system of noninteracting fermions in a steady state. We then formulated the entropy that is measured by a local observer who does not have explicit access to the scattering states but has detailed knowledge of the local spectrum and local distribution function. We find that this lack of knowledge increases the inferred entropy. Finally, we formulated the entropy that is measured without knowledge of the local distribution function but knowledge of the local mean particle number and energy. We show that the entropies formulated in the three different ways satisfy a hierarchy of inequalities with the most knowledgable formulation leading to the least entropy. We illustrated our results for a two-level system far from equilibrium. Rigorous proofs of the third law of thermodynamics were presented for generic open quantum systems which may exhibit localized states. The third law was illustrated for a benzene molecule coupled to an equilibrium reservoir in two different configurations, one exhibiting localized states and the other being a fully open quantum system. We propose an ansatz for the entropy when two-body interactions are present. We show that the ansatz is consistent with the second and third laws of thermodynamics. Based on the inequalities developed, we also use the entropy as a metric to quantify the "distance" from equilibrium. We illustrate the latter point using numerical calculations on a molecular junction driven far from equilibrium.

References

1. J. von Neumann, *Mathematical Foundations of Quantum Mechanics*. Princeton Landmarks in Mathematics and Physics (Princeton University Press, Princeton, 1996). Translation from German edition (October 28, 1996)
2. C.A. Stafford, A. Shastry, J. Chem. Phys. **146**(9), 092324 (2017). https://doi.org/10.1063/1.4975810. http://dx.doi.org/10.1063/1.4975810

3. L.D. Landau, E.M. Lifshitz, *Statistical Physics*, 3rd edn. (Butterworth-Heinemann, Oxford, 1980), pp. 160–161
4. M.A. Nielsen, I.L. Chuang, *Quantum Computation and Quantum Information: 10th Anniversary Edition*, 10th edn. (Cambridge University Press, New York, 2011)
5. M. Büttiker, Phys. Rev. Lett. **65**(23), 2901 (1990). https://doi.org/10.1103/PhysRevLett.65.2901
6. V. Gasparian, T. Christen, M. Büttiker, Phys. Rev. A **54**, 4022 (1996)
7. S. Hershfield, Phys. Rev. Lett. **70**, 2134 (1993). https://doi.org/10.1103/PhysRevLett.70.2134. https://link.aps.org/doi/10.1103/PhysRevLett.70.2134
8. G. Stefanucci, R. van Leeuwen, *Nonequilibrium Many-Body Theory Of Quantum Systems: A Modern Introduction* (Cambridge University Press, Cambridge, 2013)
9. C.A. Stafford, Phys. Rev. B **93**, 245403 (2016). https://doi.org/10.1103/PhysRevB.93.245403. http://link.aps.org/doi/10.1103/PhysRevB.93.245403
10. A. Shastry, C.A. Stafford, Phys. Rev. B **94**, 155433 (2016). https://doi.org/10.1103/PhysRevB.94.155433. http://link.aps.org/doi/10.1103/PhysRevB.94.155433
11. H. Pothier, S. Guéron, N.O. Birge, D. Esteve, M.H. Devoret, Phys. Rev. Lett. **79**, 3490 (1997). https://doi.org/10.1103/PhysRevLett.79.3490. https://link.aps.org/doi/10.1103/PhysRevLett.79.3490
12. H. Ness, Phys. Rev. B **89**, 045409 (2014). https://doi.org/10.1103/PhysRevB.89.045409. http://link.aps.org/doi/10.1103/PhysRevB.89.045409
13. J.L.W.V. Jensen, Acta Mathematica **30**(1), 175 (1906). https://doi.org/10.1007/BF02418571. http://dx.doi.org/10.1007/BF02418571
14. A. Shastry, Y. Xu, C. A. Stafford, ArXiv e-prints 1904.11628 (2019)
15. F. Reif, *Fundamentals of Statistical and Thermal Physics* (McGraw-Hill, New York, 1965)
16. M. Kolář, D. Gelbwaser-Klimovsky, R. Alicki, G. Kurizki, Phys. Rev. Lett. **109**, 090601 (2012). https://doi.org/10.1103/PhysRevLett.109.090601. https://link.aps.org/doi/10.1103/PhysRevLett.109.090601
17. B. Cleuren, B. Rutten, C. Van den Broeck, Phys. Rev. Lett. **108**, 120603 (2012). https://doi.org/10.1103/PhysRevLett.108.120603. https://link.aps.org/doi/10.1103/PhysRevLett.108.120603
18. A. Levy, R. Alicki, R. Kosloff, Phys. Rev. Lett. **109**, 248901 (2012). https://doi.org/10.1103/PhysRevLett.109.248901. https://link.aps.org/doi/10.1103/PhysRevLett.109.248901
19. A. Levy, R. Alicki, R. Kosloff, Phys. Rev. E **85**, 061126 (2012). https://doi.org/10.1103/PhysRevE.85.061126. https://link.aps.org/doi/10.1103/PhysRevE.85.061126
20. R. Kosloff, Entropy **15**(6), 2100 (2013). https://doi.org/10.3390/e15062100. http://www.mdpi.com/1099-4300/15/6/2100
21. L.A. Wu, D. Segal, P. Brumer, Sci. Rep. **3**, 1824 EP (2013). https://doi.org/10.1038/srep01824. Article
22. L. Masanes, J. Oppenheim, Nat. Commun. **8** (2017). https://doi.org/10.1038/ncomms14538
23. G.W. Ford, R.F. O'Connell, Physica E **29**, 82 (2005). https://doi.org/10.1016/j.physe.2005.05.004. https://doi.org/10.1016/j.physe.2005.05.004
24. R.F. O'Connell, J. Stat. Phys. **124**(1), 15 (2006). https://doi.org/10.1007/s10955-006-9151-6. https://doi.org/10.1007/s10955-006-9151-6
25. T.M. Nieuwenhuizen, A.E. Allahverdyan, Phys. Rev. E **66**, 036102 (2002). https://doi.org/10.1103/PhysRevE.66.036102. https://link.aps.org/doi/10.1103/PhysRevE.66.036102
26. M. Esposito, M.A. Ochoa, M. Galperin, Phys. Rev. Lett. **114**, 080602 (2015). https://doi.org/10.1103/PhysRevLett.114.080602. http://link.aps.org/doi/10.1103/PhysRevLett.114.080602
27. M.F. Ludovico, J.S. Lim, M. Moskalets, L. Arrachea, D. Sánchez, Phys. Rev. B **89**, 161306 (2014). https://doi.org/10.1103/PhysRevB.89.161306. https://link.aps.org/doi/10.1103/PhysRevB.89.161306
28. M.F. Ludovico, L. Arrachea, M. Moskalets, D. Sánchez, Phys. Rev. B **97**, 041416 (2018). https://doi.org/10.1103/PhysRevB.97.041416. https://link.aps.org/doi/10.1103/PhysRevB.97.041416

29. A. Bruch, M. Thomas, S. Viola Kusminskiy, F. von Oppen, A. Nitzan, Phys. Rev. B **93**, 115318 (2016). https://doi.org/10.1103/PhysRevB.93.115318. https://link.aps.org/doi/10.1103/PhysRevB.93.115318
30. J. Friedel, Nuovo Cimento Suppl. **7**, 287 (1958)
31. A. Bruch, C. Lewenkopf, F. von Oppen, Phys. Rev. Lett. **120**, 107701 (2018). https://doi.org/10.1103/PhysRevLett.120.107701. https://link.aps.org/doi/10.1103/PhysRevLett.120.107701
32. T. Gramespacher, M. Büttiker, Phys. Rev. B **56**, 13026 (1997). https://doi.org/10.1103/PhysRevB.56.13026. https://link.aps.org/doi/10.1103/PhysRevB.56.13026
33. J.P. Bergfield, S.M. Story, R.C. Stafford, C.A. Stafford, ACS Nano **7**(5), 4429 (2013). https://doi.org/10.1021/nn401027u

Chapter 6
Concluding Remarks

In the present book, we presented a number of interesting results pertaining to steady-state quantum fermion systems. In Chap. 2 we defined a mathematically rigorous notion of temperature and voltage for quantum systems arbitrarily far from equilibrium and having arbitrary interactions within the system. We showed that a meaningful notion of temperature and voltage for nonequilibrium systems requires the simultaneous measurement of both. This joint measurement requires that the probe be in *both* electrical *and* thermal equilibrium with the nonequilibrium system of interest. We established the notion of an ideal probe as one that operates in the broadband limit and is weakly coupled to the system of interest. The results obtained here have a deep connection with the second law of thermodynamics: We proved the uniqueness of the probe measurement and showed its close relation to the Onsager's statement of the second law of thermodynamics (which we also proved for the case of quantum thermoelectric transport). We derived necessary and sufficient conditions for the existence of a solution and found that a solution always exists, and that one may encounter negative temperature solutions if the system is driven sufficiently far away from equilibrium. We developed also the notion of entropy for steady-state noninteracting systems by first providing an expression for the exact entropy in terms of the scattering states. We then formulated the entropy inferred by a local observer with varying amounts of information regarding the system and showed that they obey a hierarchy of inequalities. The entropy was seen to be inversely related to the available information, with the most knowledgable formulation leading to the least entropy. We showed the validity of the third law of thermodynamics for open quantum systems in equilibrium and in nonequilibrium steady states. We also proposed a novel experimental method, whose working principle rests our theoretical findings, to enhance the existing spatial resolution of scanning thermal measurements by over two orders of magnitude. In addition to their theoretical merit we believe these results will have significant practical impact, for example, in characterizing nonequilibrium device performances.

© Springer Nature Switzerland AG 2019
A. Shastry, *Theory of Thermodynamic Measurements of Quantum Systems Far from Equilibrium*, Springer Theses, https://doi.org/10.1007/978-3-030-33574-8_6

The work presented here leaves plenty of interesting open problems. For example, we did not consider the case where the fermionic spin degeneracy is broken such as in magnetic systems. The first law of thermodynamics would then have an additional magnetic term and it is unclear whether the additional thermodynamic variable (magnetic moment) entering the first law can be defined uniquely outside equilibrium. Furthermore, we did not consider thermodynamic measurements of the bosonic degrees of freedom which is an interesting question to pursue. Notions of nonequilibrium entropy were presented rigorously in the absence of interactions and, though we provide an ansatz in the presence of interactions, it appears to be a formidable theoretical challenge when two-body interactions are present. Although we commented upon the relationship of our notion of entropy to the entanglement entropy, we did not elaborate on it. All of our analyses in this book pertained to nonequilibrium steady states. It is entirely unclear how one might conceive of these thermodynamic notions in the realm of time-varying nonequilibrium states.

From a broader perspective, nonequilibrium thermodynamics is a rapidly developing field of study and it is certainly a very exciting time for theoretical work. Open quantum systems have been studied from a dynamical viewpoint [1] (where system + environment undergoes unitary evolution but the system dynamics described by the reduced density matrix is nonunitary), e.g., using Master equation approaches, and the notion of decoherence has had considerable success in explaining the quantum to classical transition. The elucidation of how thermodynamic equilibrium states emerge from this underlying process of decoherence is one of the aims of quantum thermodynamics and has seen a flurry of activity in recent years [2]. There are also approaches to nonequilibrium thermodynamics which take ideas from quantum information theory where thermal nonequilibrium states are seen as "resources" [3]; for example, such resources may be consumed as fuel to achieve an erasure operation which irreversibly generates entropy. There is also a great impetus on the experimental side, for example, a recent experiment [4] employed two initially quantum-correlated spins (a resource) to reverse the thermodynamic arrow of time, i.e., having heat flow from the colder system to the hotter one; this of course is not a violation of the Clausius statement since the heat flow is not the "sole effect." Stochastic thermodynamics [5] is a relatively recent development which aims to extend notions of classical irreversible thermodynamics to the level of individual particle trajectories and is now being increasingly employed to study nonequilibrium quantum systems, e.g., in the context of the irreversibility brought about by a quantum measurement [6]. In summary, there is a great ongoing effort to understand some of the most fundamental aspects of quantum mechanics and thermodynamics. The confluence of these many different approaches, including the Green's function approach, to nonequilibrium thermodynamics of quantum systems is sure to create many fruitful discussions in the years to come.

References

1. H.P. Breuer, F. Petruccione, *The Theory of Open Quantum Systems* (Oxford University Press, Oxford, 2002)
2. J. Gemmer, M. Michel, G. Mahler, *Quantum Thermodynamics: Emergence of Thermodynamic Behavior Within Composite Quantum Systems* (Springer, Berlin, 2009). https://doi.org/10.1007/978-3-540-70510-9
3. G. Gour, M.P. Mller, V. Narasimhachar, R.W. Spekkens, N.Y. Halpern, Phys. Rep. **583**, 1 (2015). http://dx.doi.org/10.1016/j.physrep.2015.04.003. http://www.sciencedirect.com/science/article/pii/S037015731500229X
4. K. Micadei, J. Peterson, A. Souza, R. Sarthour, I. Oliveria, G. Landi, T. Batalhao, R. Serra, E. Lutz, arXiv eprints (2017). https://doi.org/arXiv:1711.0332. https://arxiv.org/abs/1711.03323
5. U. Seifert, Rep. Prog. Phys. **75**(12), 126001 (2012). http://stacks.iop.org/0034-4885/75/i=12/a=126001
6. C. Elouard, D.A. Herrera-Martí, M. Clusel, A. Auffèves, npj Quantum Inf. **3**(1), 9 (2017). https://doi.org/10.1038/s41534-017-0008-4

Appendix A
The Nonequilibrium Steady State

We consider a system whose Hamiltonian $\hat{H}$ is independent of time, but is driven out of equilibrium, e.g., by electrical and/or thermal bias. The nonequilibrium steady state is described by a density matrix $\hat{\rho}$ that is time-independent. The expectation values of observables are given by their usual prescription in statistical physics

$$\langle \hat{Q} \rangle = \mathrm{Tr}\left\{ \hat{\rho}\hat{Q} \right\} = \sum_{\mu,\nu} \rho_{\mu\nu} \langle \nu | \hat{Q} | \mu \rangle. \tag{A.1}$$

The retarded and advanced Green's functions are defined as follows:[1]

$$G^r_{\alpha\beta}(t) = -i\theta(t)\langle \{d_\alpha(t), d^\dagger_\beta(0)\}\rangle, \tag{A.2}$$

while its Hermitian conjugate is

$$G^a_{\alpha\beta}(t) = i\theta(-t)\langle \{d_\alpha(t), d^\dagger_\beta(0)\}\rangle, \tag{A.3}$$

where

$$d_\alpha(t) = e^{i\frac{\hat{H}}{\hbar}t} d_\alpha(0) e^{-i\frac{\hat{H}}{\hbar}t} \tag{A.4}$$

evolves according to the Heisenberg equation of motion for a system with Hamiltonian $\hat{H}$. Here, α, β denote basis states in the one-body Hilbert space of the system. The retarded and advanced Green's functions describe the particle and hole propagation within the system.

---

[1]In this appendix, the operators acting on the full many-body Hilbert space (also called Fock space) are denoted by a "hat" whereas the operators acting on the one-body Hilbert space of the system are denoted without a hat. The creation and annihilation operators are, however, also denoted without the hat since it is clear that they indeed act on the Fock space.

© Springer Nature Switzerland AG 2019
A. Shastry, *Theory of Thermodynamic Measurements of Quantum Systems Far from Equilibrium*, Springer Theses, https://doi.org/10.1007/978-3-030-33574-8

The "lesser" and "greater" Green's functions [1] used in this book are defined as follows:

$$G^<_{\alpha\beta}(t) \equiv i \langle d^\dagger_\beta(0) d_\alpha(t) \rangle, \tag{A.5}$$

while its Hermitian conjugate is

$$G^>_{\alpha\beta}(t) \equiv -i \langle d_\alpha(t) d^\dagger_\beta(0) \rangle. \tag{A.6}$$

The lesser and greater Green's functions describe the (nonequilibrium) particle and hole occupancies within the system.

The four Green's functions introduced above obey the general relation

$$G^r - G^a = G^> - G^<. \tag{A.7}$$

The spectral representation uses the eigenbasis of the Hamiltonian $\hat{H}|\nu\rangle = E_\nu|\nu\rangle$, where ν denotes a many-body energy eigenstate. One may write the "lesser" Green's function as

$$G^<_{\alpha\beta}(\omega) = 2\pi i \sum_{\mu,\mu',\nu} \rho_{\mu\nu} \langle \nu | d^\dagger_\beta | \mu' \rangle \langle \mu' | d_\alpha | \mu \rangle$$
$$\times \delta\left(\omega - \frac{E_\mu - E_{\mu'}}{\hbar}\right), \tag{A.8}$$

while the "greater" Green's function becomes

$$G^>_{\alpha\beta}(\omega) = -2\pi i \sum_{\mu,\mu',\nu} \rho_{\mu\nu} \langle \nu | d_\alpha | \mu' \rangle \langle \mu' | d^\dagger_\beta | \mu \rangle$$
$$\times \delta\left(\omega - \frac{E_{\mu'} - E_\nu}{\hbar}\right). \tag{A.9}$$

The spectral function $A(\omega)$ is given by

$$A(\omega) \equiv \frac{1}{2\pi i}\left(G^<(\omega) - G^>(\omega)\right), \tag{A.10}$$

and can be expressed in the spectral representation as

$$A_{\alpha\beta}(\omega) = \sum_{\mu,\mu',\nu} \left[\rho_{\mu\nu} \langle \nu | d^\dagger_\beta | \mu' \rangle \langle \mu' | d_\alpha | \mu \rangle + \rho_{\nu\mu'} \langle \mu' | d_\alpha | \mu \rangle \langle \mu | d^\dagger_\beta | \nu \rangle \right]$$
$$\times \delta\left(\omega - \frac{E_\mu - E_{\mu'}}{\hbar}\right). \tag{A.11}$$

A.1   Sum Rule for the Spectral Function

Equation (A.11) leads to the following sum rule for the spectral function:

$$
\int_{-\infty}^{\infty} d\omega A_{\alpha\beta}(\omega) = \sum_{\mu,\nu} \rho_{\mu\nu}\langle\nu|d_{\beta}^{\dagger}d_{\alpha}|\mu\rangle + \sum_{\mu',\nu} \rho_{\nu\mu'}\langle\mu'|d_{\alpha}d_{\beta}^{\dagger}|\nu\rangle
$$

$$
= \sum_{\mu,\nu} \rho_{\mu\nu}\langle\nu|d_{\beta}^{\dagger}d_{\alpha} + d_{\alpha}d_{\beta}^{\dagger}|\mu\rangle
$$

$$
= \sum_{\mu,\nu} \rho_{\mu\nu}\langle\nu|\delta_{\alpha\beta}|\mu\rangle \tag{A.12}
$$

$$
= \sum_{\mu,\nu} \rho_{\mu\nu}\delta_{\mu\nu}\delta_{\alpha\beta}
$$

$$
= \delta_{\alpha\beta}\,\mathrm{Tr}\{\hat{\rho}\}
$$

$$
= \delta_{\alpha\beta}.
$$

In our theory of local thermodynamic measurements, the quantity of interest is the local spectrum of the system sampled by the probe $\bar{A}(\omega)$, defined in Eq. (2.55). This obeys a further sum rule in the broadband limit (*ideal probe*), discussed below.

Local Spectrum in the Broadband Limit

The probe-system coupling is energy independent in the broadband limit, $\Gamma^{P}(\omega) =$ const, and we write $\mathrm{Tr}\{\Gamma^{P}\} = \bar{\Gamma}^{P}$ for its trace. The local spectrum sampled by the probe $\bar{A}(\omega)$ defined in Eq. (2.55) can be written in the broadband limit as

$$
\bar{A}(\omega) = \frac{1}{\bar{\Gamma}^{P}} \sum_{\alpha,\beta} \langle\beta|\Gamma^{P}|\alpha\rangle A_{\alpha\beta}(\omega). \tag{A.13}
$$

In this limit, it obeys a further sum rule:

$$
\int_{-\infty}^{\infty} d\omega \bar{A}(\omega) = \frac{1}{\bar{\Gamma}^{P}} \sum_{\alpha,\beta} \langle\beta|\Gamma^{P}|\alpha\rangle \int_{-\infty}^{\infty} d\omega A_{\alpha\beta}(\omega)
$$

$$
= \frac{1}{\bar{\Gamma}^{P}} \sum_{\alpha,\beta} \langle\beta|\Gamma^{P}|\alpha\rangle\delta_{\alpha\beta} \tag{A.14}
$$

$$
= 1.
$$

The broadband limit is special in that the measurement is determined by the local properties of the system itself, and is not influenced by the spectrum of the probe. In this limit, the local spectrum $\bar{A}(\omega)$ obeys the sum rule (A.14) since the probe samples the same subsystem at all energies. One should not expect such a local sum rule to hold outside the broadband limit, since the probe samples different subsystems at different energies.

A.2   Diagonality of $\hat{\rho}$

We have, for any observable $\hat{Q}$,

$$
\begin{aligned}
\langle \hat{Q}(t) \rangle &= \sum_{\mu,\nu} \rho_{\mu\nu} \langle \nu | \hat{Q}(t) | \mu \rangle \\
&= \sum_{\mu,\nu} \rho_{\mu\nu} \langle \nu | e^{i \frac{\hat{H}}{\hbar} t} \hat{Q} e^{-i \frac{\hat{H}}{\hbar} t} | \mu \rangle \\
&= \sum_{\mu,\nu} \rho_{\mu\nu} e^{-i \frac{E_\mu - E_\nu}{\hbar} t} \langle \nu | \hat{Q} | \mu \rangle.
\end{aligned}
\tag{A.15}
$$

The system observables must be independent of time in steady state. Therefore $\hat{\rho}$ must be diagonal in the energy basis, as seen from the above equation. The nondiagonal parts of $\hat{\rho}$ in the energy basis, when they exist, must be in a degenerate subspace so that $E_\mu = E_\nu$ in the above equation.

For states degenerate in energy, the boundary conditions determining the nonequilibrium steady state will determine the basis in which $\hat{\rho}$ is diagonal. Henceforth, we work in that basis.

A.3   Positivity of $-i G^<(\omega)$ and $i G^>(\omega)$

Working in the energy eigenbasis in which $\hat{\rho}$ is diagonal,

$$
\begin{aligned}
-i \langle \alpha | G^<(\omega) | \alpha \rangle &\equiv -i G^<_{\alpha\alpha}(\omega) \\
&= 2\pi \sum_{\mu,\mu'} \rho_{\mu\mu} \left| \langle \mu | d_\alpha^\dagger | \mu' \rangle \right|^2 \delta\left(\omega - \frac{E_\mu - E_{\mu'}}{\hbar} \right) \geq 0.
\end{aligned}
\tag{A.16}
$$

Similarly,

$$
\begin{aligned}
i \langle \alpha | G^>(\omega) | \alpha \rangle &\equiv i G^>_{\alpha\alpha}(\omega) \\
&= 2\pi \sum_{\mu,\mu'} \rho_{\mu\mu} \left| \langle \mu | d_\alpha^\dagger | \mu' \rangle \right|^2 \delta\left(\omega - \frac{E_{\mu'} - E_\mu}{\hbar} \right) \geq 0.
\end{aligned}
\tag{A.17}
$$

It follows that

$$\langle \alpha | A(\omega) | \alpha \rangle = \frac{1}{2\pi} \langle \alpha | -iG^<(\omega) + iG^>(\omega) | \alpha \rangle \geq 0. \quad (A.18)$$

Therefore, all three operators $-iG^<(\omega)$, $iG^>(\omega)$, and $A(\omega)$ are positive-semidefinite.

A.4 $0 \leq f_s(\omega) \leq 1$

The nonequilibrium distribution function $f_s(\omega)$ was defined in Eq. (2.3) as

$$f_s(\omega) \equiv \frac{\mathrm{Tr}\{\Gamma^P(\omega)G^<(\omega)\}}{2\pi i\, \mathrm{Tr}\{\Gamma^P(\omega)A(\omega)\}}. \quad (A.19)$$

We have $\Gamma^P(\omega) > 0$ by causality [1]:

$$\mathrm{Im}\, \Sigma_p^r(\omega) = -\frac{1}{2}\Gamma^P(\omega) < 0. \quad (A.20)$$

Let $\Gamma^P |\gamma_p\rangle = \gamma_p |\gamma_p\rangle$, where $\gamma_p \geq 0$ and some γ_p satisfy $\gamma_p > 0$. The energy dependence is taken to be implicit. The traces in Eq. (A.19) may be evaluated in the eigenbasis of Γ^P, yielding:

$$\begin{aligned}
f_s(\omega) &= \frac{\sum_{\gamma_p} \gamma_p \langle \gamma_p | G^<(\omega) | \gamma_p \rangle}{2\pi i \sum_{\gamma_p} \gamma_p \langle \gamma_p | A(\omega) | \gamma_p \rangle} \\
&= \frac{\sum_{\gamma_p} \gamma_p \langle \gamma_p | -iG^<(\omega) | \gamma_p \rangle}{\sum_{\gamma_p} \gamma_p \langle \gamma_p | -iG^<(\omega) + iG^>(\omega) | \gamma_p \rangle}.
\end{aligned} \quad (A.21)$$

Therefore

$$0 \leq f_s(\omega) \leq 1. \quad (A.22)$$

Reference

1. G. Stefanucci, R. van Leeuwen, *Nonequilibrium Many-Body Theory of Quantum Systems: A Modern Introduction* (Cambridge University Press, Cambridge, 2013)

Appendix B
Noninvasive Probes

Our main results in Chap. 2 relied upon the assumption of a noninvasive probe. We explained the physical basis for this assumption in Sect. 2.3.2, and we understood it to mean that the local probe-system transmission function $\mathcal{T}_{ps}$ and the local nonequilibrium distribution function f_s are independent of the probe bias parameters (μ_p, T_p). In this appendix, we clarify the implicit mathematical details that have gone into this assumption of a noninvasive probe.

f_s and $\mathcal{T}_{ps}$ have been defined in Eqs. (2.3) and (2.4), respectively, and they depend upon the Green's functions of the nonequilibrium quantum system. The Green's functions of the system do depend upon the probe parameters (μ_p, T_p) and we clarify this dependence. We label the probe parameter simply as $x_p \in \{\mu_p, T_p\}$, which can be taken to mean either the chemical potential or the temperature of the probe.

In order to characterize the noninvasive probe limit, we introduce a dimensionless parameter λ, and write the probe-system coupling as $\Gamma^p(\omega) = \lambda \tilde{\Gamma}^p(\omega)$. Without loss of generality, we may set $\text{Tr}\left\{\tilde{\Gamma}^p(\mu_0)\right\} = \sum_{\alpha \neq p} \text{Tr}\{\Gamma^\alpha(\mu_0)\}$, where $\Gamma^\alpha(\omega)$ is the tunneling-width matrix describing the coupling of lead α (e.g., source, drain, etc.) to the system, and μ_0 is the equilibrium chemical potential of the system (or some other convenient reference value). The parameter

$$\lambda \equiv \frac{\text{Tr}\{\Gamma^p(\mu_0)\}}{\sum_{\alpha \neq p} \text{Tr}\{\Gamma^\alpha(\mu_0)\}} \ll 1 \tag{B.1}$$

thus gives the condition for a weakly coupled probe.

The currents flowing into the probe from the system are given by Eq. (2.2) as

$$I_p^{(\nu)} = \frac{1}{h} \int_{-\infty}^{\infty} d\omega (\omega - \mu_p)^\nu \mathcal{T}_{ps}(\omega)[f_s(\omega) - f_p(\omega)], \tag{B.2}$$

© Springer Nature Switzerland AG 2019
A. Shastry, *Theory of Thermodynamic Measurements of Quantum Systems Far from Equilibrium*, Springer Theses, https://doi.org/10.1007/978-3-030-33574-8

where

$$\mathcal{T}_{ps} = \lambda 2\pi \, \mathrm{Tr}\Big\{\tilde{\Gamma}^p A\Big\}$$

$$= \lambda 2\pi \, \mathrm{Tr}\Big\{\tilde{\Gamma}^p A|_{\lambda=0}\Big\} + \lambda^2 2\pi \, \mathrm{Tr}\Big\{\tilde{\Gamma}^p \frac{\partial A}{\partial \lambda}\Big|_{\lambda=0}\Big\}$$

$$+ \mathcal{O}(\lambda^3) \tag{B.3}$$

and

$$\mathcal{T}_{ps} f_s = -i\lambda \, \mathrm{Tr}\Big\{\tilde{\Gamma}^p G^<\Big\}$$

$$= -i\lambda \, \mathrm{Tr}\Big\{\tilde{\Gamma}^p G^<|_{\lambda=0}\Big\} - i\lambda^2 \, \mathrm{Tr}\Big\{\tilde{\Gamma}^p \frac{\partial G^<}{\partial \lambda}\Big|_{\lambda=0}\Big\}$$

$$+ \mathcal{O}(\lambda^3) \tag{B.4}$$

[cf. Eqs. (2.3) and (2.4)]. From Eqs. (B.2) to (B.4), we see that $I_p^{(\nu)} \sim \mathcal{O}(\lambda)$. Similarly, it can be shown that

$$\frac{\partial I_p^{(\nu)}}{\partial x_p} = -I_p^{(0)} \delta_{\nu,1} \delta_{x_p,\mu_p}$$

$$- \frac{\lambda}{\hbar} \int_{-\infty}^{\infty} d\omega (\omega - \mu_p)^{\nu} \, \mathrm{Tr}\Big\{\tilde{\Gamma}^p A|_{\lambda=0}\Big\} \frac{\partial f_p}{\partial x_p}$$

$$+ \mathcal{O}(\lambda^2). \tag{B.5}$$

The leading-order results for the gradients are also $\mathcal{O}(\lambda)$, and agree with Eqs. (2.16) and (2.17). The noninvasive probe limit consists in keeping only the terms $\mathcal{O}(\lambda)$ in Eqs. (B.3)–(B.5), and underlies the analysis presented in the body of the article. Deviations from the noninvasive probe limit appear as terms $\mathcal{O}(\lambda^2)$ and higher, which we now proceed to derive.

B.1 Dependence of G on λ and x_p

Standard NEGF arguments can be used to elucidate the dependence of the system Green's functions on λ and x_p. Let G_0 denote the Green's function of the isolated quantum system without two-body interactions, and let Σ denote the self-energy describing two-body interactions and coupling to various reservoirs, including the probe. Dyson's equation for the retarded (advanced) Green's function is [1]

$$G^{r,a} = G_0^{r,a} + G_0^{r,a} \Sigma^{r,a} G^{r,a}. \tag{B.6}$$

The Keldysh equation for $G^<$ is [1]

$$G^<(\omega) = G^r(\omega)\Sigma^<(\omega)G^a(\omega), \tag{B.7}$$

where the "lesser" self-energy is

$$\Sigma^< = i\lambda\tilde{\Gamma}^p(\omega)f_p(\omega) + i\sum_{\alpha\neq p}\Gamma^\alpha(\omega)f_u(\omega) + \Sigma^<_{\text{int}}, \tag{B.8}$$

and $\Sigma^<_{\text{int}}$ is the self-energy contribution due to electron–electron, electron–phonon, electron–photon interactions, etc. Similarly, the spectral function A may be expressed as

$$2\pi A(\omega) = G^r(\omega)\Gamma(\omega)G^a(\omega), \tag{B.9}$$

where

$$\Gamma(\omega) = \lambda\tilde{\Gamma}^p(\omega) + \sum_{\alpha\neq p}\Gamma^\alpha(\omega) + \Gamma_{\text{int}}(\omega), \tag{B.10}$$

and $\Gamma_{\text{int}} = i(\Sigma^r_{\text{int}} - \Sigma^a_{\text{int}})$ is the contribution due to two-body interactions. Note that all the terms appearing on the r.h.s. of Eq. (B.10) are positive definite due to causality.

Differentiating the self-energies with respect to x_p gives

$$\frac{\partial\Sigma^<}{\partial x_p} = i\lambda\tilde{\Gamma}^p\frac{\partial f_p}{\partial x_p} + \frac{\partial\Sigma^<_{\text{int}}}{\partial x_p}, \tag{B.11}$$

and

$$\frac{\partial\Sigma^{r,a}}{\partial x_p} = \frac{\partial\Sigma^{r,a}_{\text{int}}}{\partial x_p}. \tag{B.12}$$

Using Eqs. (B.6), (B.7), (B.11), and (B.12), it can be shown that

$$\frac{\partial G^{r,a}}{\partial x_p} = G^{r,a}\frac{\partial\Sigma^{r,a}_{\text{int}}}{\partial x_p}G^{r,a}, \tag{B.13}$$

$$\frac{\partial G^<}{\partial x_p} = i\lambda G^r\tilde{\Gamma}^p G^a\frac{\partial f_p}{\partial x_p} + G^r\frac{\partial\Sigma^<_{\text{int}}}{\partial x_p}G^a$$
$$+ G^r\frac{\partial\Sigma^r_{\text{int}}}{\partial x_p}G^< + G^<\frac{\partial\Sigma^a_{\text{int}}}{\partial x_p}G^a. \tag{B.14}$$

Using $2\pi iA = G^a - G^r$, the derivative of the spectral function may be written as

$$2\pi i \frac{\partial A(\omega)}{\partial x_p} = G^a \frac{\partial \Sigma_{\text{int}}^a}{\partial x_p} G^a - G^r \frac{\partial \Sigma_{\text{int}}^r}{\partial x_p} G^r. \tag{B.15}$$

Finally, the derivatives of Σ_{int} are given by

$$\frac{\partial \Sigma_{\text{int}}^\gamma(\omega)}{\partial x_p} = \sum_{\eta=r,a,<} \int_{-\infty}^\infty d\omega' K^{\gamma\eta}(\omega, \omega') \frac{\partial G^\eta(\omega')}{\partial x_p}, \tag{B.16}$$

where

$$K^{\gamma\eta}(\omega, \omega') \equiv \frac{\delta \Sigma_{\text{int}}^\gamma(\omega)}{\delta G^\eta(\omega')} \tag{B.17}$$

is the irreducible kernel for the 2-particle Green's function [1].

Equations (B.13), (B.14), and (B.16) are three coupled linear (integral) equations for $\partial G/\partial x_p$ and $\partial \Sigma_{\text{int}}/\partial x_p$. The only inhomogeneous term [first term on the r.h.s. of Eq. (B.14)] is $\mathcal{O}(\lambda)$. Let

$$\frac{\partial G^\gamma(\omega)}{\partial x_p} \equiv \lambda F_{x_p}^\gamma(\omega), \tag{B.18}$$

$$\frac{\partial \Sigma_{\text{int}}^\gamma(\omega)}{\partial x_p} \equiv \lambda S_{x_p}^\gamma(\omega). \tag{B.19}$$

F and S satisfy the equations

$$F_{x_p}^{r,a} = G^{r,a} S_{x_p}^{r,a} G^{r,a}, \tag{B.20}$$

$$F_{x_p}^< = i G^r \tilde{\Gamma}^p G^a \frac{\partial f_p}{\partial x_p} + G^r S_{x_p}^< G^a$$
$$+ G^r S_{x_p}^r G^< + G^< S_{x_p}^a G^a, \tag{B.21}$$

and

$$S_{x_p}^\gamma = \sum_{\eta=r,a,<} K^{\gamma\eta} F_{x_p}^\eta, \tag{B.22}$$

where the energy integral on the r.h.s. of Eq. (B.22) is implicit. The leading-order solution is obtained by setting $G^\gamma = G^\gamma|_{\lambda=0}$ in Eqs. (B.20) and (B.21), so we see that $\partial G/\partial x_p$, $\partial \Sigma_{\text{int}}/\partial x_p \sim \mathcal{O}(\lambda)$, and can be neglected in the noninvasive probe limit. There exist a number of additional terms in $\partial G/\partial \lambda|_{\lambda=0}$ that are independent of x_p, but these do not affect the proofs of Theorems 2.1–2.3.

B.2   Proof of Uniqueness

We are now in a position to evaluate the dependence of the currents $I_p^{(\nu)}$ on $x_p \in \{\mu_p, T_p\}$. Taking the derivative of Eq. (B.2) using the results of Appendix B.1, one obtains the exact expression

$$
\frac{\partial I_p^{(\nu)}}{\partial x_p} = -I_p^{(0)} \delta_{\nu,1} \delta_{x_p, \mu_p} - \frac{\lambda}{h} \int_{-\infty}^{\infty} d\omega (\omega - \mu_p)^{\nu} \operatorname{Tr}\left\{ \tilde{\Gamma}^p G^r (\Gamma - \Gamma^p) G^a \right\} \frac{\partial f_p}{\partial x_p}
$$

$$
- \frac{i\lambda^2}{h} \int_{-\infty}^{\infty} d\omega (\omega - \mu_p)^{\nu} \operatorname{Tr}\left\{ \tilde{\Gamma}^p \left(G^r S_{x_p}^< G^a + G^r S_{x_p}^r G^< + G^< S_{x_p}^a G^a \right) \right\}
$$

$$
+ \frac{i\lambda^2}{h} \int_{-\infty}^{\infty} d\omega (\omega - \mu_p)^{\nu} f_p(\omega) \operatorname{Tr}\left\{ \tilde{\Gamma}^p \left(G^a S_{x_p}^a G^a - G^r S_{x_p}^r G^r \right) \right\}.
$$

$$\tag{B.23}$$

To leading order in λ, Eq. (B.23) reduces to the result given in Eqs. (2.16) and (2.17), while the corrections are $\mathcal{O}(\lambda^2)$ or higher. Thus Theorems 2.1 and 2.2 hold to leading order in λ for systems with arbitrary two-body interactions, and the noninvasive probe limit may be precisely defined as the limit $\lambda \ll 1$.

Special Case: Noninteracting System
Without two-body interactions, only the first two terms in Eq. (B.23) survive. The current gradients therefore have the same form as in Eqs. (2.16) and (2.17), while the $\mathcal{L}_{ps}^{(\nu)}$ coefficients have the same form [cf. Eq. (2.18)] but with the transmission function replaced by

$$
\tilde{\mathcal{T}}_{ps}(\omega) = \operatorname{Tr}\left\{ \Gamma^p(\omega) G^r(\omega) \left(\Gamma(\omega) - \Gamma^p(\omega) \right) G^a(\omega) \right\},
\tag{B.24}
$$

which is positive due to causality [see Eq. (B.10)]. Theorem 2.1 therefore still holds. The uniqueness result as stated in Theorem 2.2 holds also, since the argument only makes use of current gradients. We note that Theorems 2.1 and 2.2 hold for arbitrarily strong probe couplings when two-body interactions are absent.

Example: Hartree–Fock Approximation
In the Hartree–Fock approximation, the irreducible kernel defined in Eq. (B.17) has the form

$$
K^{r<} \equiv \frac{\delta \left(\Sigma_{\mathrm{HF}}^r \right)_{nm}}{\delta \left(G^< \right)_{ij}} = U_{nj} \delta_{nm} \delta_{ij} - U_{nm} \delta_{ni} \delta_{mj},
\tag{B.25}
$$

where U_{nm} is the Coulomb integral between orthonormal basis orbitals n and m of the system. Furthermore, $K^{a<} = K^{r<}$ and $K^{<<} = K^{><} = 0$. Equation (B.22) therefore reduces to $S_{x_p}^< = 0$ and

$$\left(S_{x_p}^{r,a}\right)_{nm} = \delta_{nm} \sum_j U_{nj} \int_{-\infty}^{\infty} d\omega' \left(F_{x_p}^<(\omega')\right)_{jj}$$

$$-U_{nm} \int_{-\infty}^{\infty} d\omega' \left(F_{x_p}^<(\omega')\right)_{nm}. \tag{B.26}$$

B.3   Proof of Existence

The proof of Theorem 2.3 is based on an analysis of the quantities $\langle \dot{N} \rangle|_{f_s}$, $\langle \dot{N} \rangle|_{f_p}$, $\langle \dot{E} \rangle|_{f_s}$, and $\langle \dot{E} \rangle|_{f_p}$ defined in Eqs. (2.38)–(2.41), respectively. These quantities are simply energy integrals of $\omega^\nu \mathcal{T}_{ps} f_s$ and $\omega^\nu \mathcal{T}_{ps} f_p$, with $\nu = 0, 1$, whose dependence on the small parameter λ is given in Eqs. (B.3) and (B.4). Keeping only terms $\mathcal{O}(\lambda)$ (noninvasive probe limit), these quantities reduce to the form considered in Sect. 2.5, so that Theorem 2.3 and Corollary 2.3.1 hold as before. Deviations from the noninvasive probe limit involve corrections $\mathcal{O}(\lambda^2)$ and higher, and it is an open question whether a unique solution to the probe equilibration conditions (2.42) exists for arbitrarily strong probe-system coupling in the presence of interactions.

Special Case: Noninteracting System
For systems without two-body interactions, the proof of Theorem 2.3 can be straightforwardly extended to the case of arbitrarily strong probe-system coupling. Using Eqs. (B.3) and (B.10) with $\Gamma_{\text{int}} = 0$, one can write

$$\mathcal{T}_{ps} = \sum_\alpha \text{Tr}\{\Gamma^p G^r \Gamma^\alpha G^a\}. \tag{B.27}$$

Similarly, using Eqs. (B.4) and (B.8) with $\Sigma_{\text{int}}^< = 0$, one has

$$\mathcal{T}_{ps} f_s = \sum_\alpha \text{Tr}\{\Gamma^p G^r \Gamma^\alpha G^a\} f_\alpha. \tag{B.28}$$

The probe equilibration conditions (2.42) whose solution we seek may be rewritten

$$\langle \dot{N} \rangle|_{f_s} - \langle \dot{N} \rangle|_{f_p} = 0,$$

$$\langle \dot{E} \rangle|_{f_s} - \langle \dot{E} \rangle|_{f_p} = 0. \tag{B.29}$$

The integrands in both conditions involve

$$\mathcal{T}_{ps}[f_s - f_p] = \sum_{\alpha \neq p} \text{Tr}\{\Gamma^p G^r \Gamma^\alpha G^a\}[f_\alpha - f_p]$$

$$= \tilde{\mathcal{T}}_{ps}[\tilde{f}_s - f_p], \tag{B.30}$$

where $\tilde{\mathcal{T}}_{ps}$ is given by Eq. (B.24) and

$$\tilde{f}_s = \frac{\sum_{\alpha \neq p} \mathrm{Tr}\{\Gamma^p G^r \Gamma^\alpha G^a\} f_\alpha}{\sum_{\alpha \neq p} \mathrm{Tr}\{\Gamma^p G^r \Gamma^\alpha G^a\}}. \tag{B.31}$$

$\tilde{f}_s$ and $\tilde{\mathcal{T}}_{ps}$ are both independent of x_p for the noninteracting system, and $\langle \dot{N} \rangle|_{f_s}$, $\langle \dot{N} \rangle|_{f_p}$, $\langle \dot{E} \rangle|_{f_s}$, and $\langle \dot{E} \rangle|_{f_p}$ can be redefined using $\tilde{f}_s$ and $\tilde{\mathcal{T}}_{ps}$ without affecting the conditions (B.29). Therefore the proofs of Theorem 2.3 and Corollary 2.3.1 hold for arbitrarily strong probe-system coupling in systems without two-body interactions.

Reference

1. G. Stefanucci, R. van Leeuwen, *Nonequilibrium Many-Body Theory of Quantum Systems: A Modern Introduction* (Cambridge University Press, Cambridge, 2013)

Appendix C
Elastic Transport

C.1 Nonequilibrium Distribution Function

We derive the form of the nonequilibrium distribution function $f_s(\omega)$ when the transport is dominated by elastic processes. We assume a nanostructure connected to M reservoirs, including the probe. Equation (2.2) takes the form of Eq. (3.14) when the transport is elastic, and we have

$$2\pi \, \mathrm{Tr}\big\{\Gamma^p(\omega)A(\omega)\big\}\big(f_s(\omega) - f_p(\omega)\big)$$
$$= \sum_{\alpha=1}^{M} \mathcal{T}_{p\alpha}(\omega)\big(f_\alpha(\omega) - f_p(\omega)\big). \tag{C.1}$$

Now, we wish to rewrite the above equation in terms of the local properties sampled by the probe:

$$\frac{\mathrm{Tr}\{\Gamma^p(\omega)A(\omega)\}}{\mathrm{Tr}\{\Gamma^p(\omega)\}}\big(f_s(\omega) - f_p(\omega)\big)$$
$$= \sum_{\alpha=1}^{M} \frac{\mathrm{Tr}\{\Gamma^p(\omega)G^r(\omega)\Gamma^\alpha(\omega)G^a(\omega)\}}{2\pi \, \mathrm{Tr}\{\Gamma^p(\omega)\}}\big(f_\alpha(\omega) - f_p(\omega)\big), \tag{C.2}$$

where the first factor on the l.h.s. is the mean local spectrum $\bar{A}(\omega)$ sampled by the probe, defined by Eq. (3.5), and we used Eq. (C.13) for the elastic transmissions on the r.h.s. We define the injectivity of a reservoir α sampled by the probe as

$$\rho_{p\alpha}(\omega) = \frac{1}{2\pi} \frac{\mathrm{Tr}\{\Gamma^p(\omega)G^r(\omega)\Gamma^\alpha(\omega)G^a(\omega)\}}{\mathrm{Tr}\{\Gamma^p(\omega)\}}, \tag{C.3}$$

© Springer Nature Switzerland AG 2019

A. Shastry, *Theory of Thermodynamic Measurements of Quantum Systems Far from Equilibrium*, Springer Theses, https://doi.org/10.1007/978-3-030-33574-8

for the factors appearing on the r.h.s. of Eq. (C.2). Injectivity of a reservoir α has been previously defined [1] as the local partial density of states (LPDOS) associated with the electrons originating from reservoir α and, due to number conservation, the sum of injectivities of the reservoirs gives the local density of states (LDOS). We state an equivalent result for the injectivities defined in Eq. (C.3) in the following paragraph. Before proceeding, we note that the injectivities sampled by the probe, in Eq. (C.3), reduces to the LPDOS for electrons injected by reservoir α when the probe coupling is maximally local, i.e., $[\Gamma^p(\omega)]_{ij} = \Gamma^p(\omega)\delta_{in}\delta_{jn}$ and becomes essentially independent of the probe coupling when it is weak. Equation (C.3) also extends to $\alpha = p$ and defines the probe injectivity sampled by itself, which becomes negligible in the limit of weak coupling.

It can be shown that the spectrum can be written as [2]

$$A(\omega) = \frac{1}{2\pi} G^r(\omega)\Gamma(\omega)G^a(\omega), \tag{C.4}$$

where $\Gamma(\omega)$ is given by

$$\Gamma(\omega) = \sum_\alpha \Gamma^\alpha(\omega). \tag{C.5}$$

The contribution due to interactions $\Gamma^{\text{int}}(\omega)$ in Eq. (C.5) is missing since the interaction self-energy is Hermitian for elastic processes. Equations (C.3)–(C.5) imply:

$$\sum_{\alpha=1}^{M} \rho_{p\alpha}(\omega) = \bar{A}(\omega). \tag{C.6}$$

From Eq. (C.2), we write

$$\bar{A}(\omega)\big(f_s(\omega) - f_p(\omega)\big) = \sum_{\alpha=1}^{M} \rho_{p\alpha}(\omega)\big(f_\alpha(\omega) - f_p(\omega)\big) \tag{C.7}$$

and Eq. (C.6) implies

$$\bar{A}(\omega)f_s(\omega) = \sum_{\alpha=1}^{M} \rho_{p\alpha}(\omega)f_\alpha(\omega). \tag{C.8}$$

Finally, $f_s(\omega)$ can be written as

$$f_s(\omega) = \sum_{\alpha=1}^{M} \frac{\rho_{p\alpha}(\omega)}{\bar{A}(\omega)} f_\alpha(\omega). \tag{C.9}$$

$$\therefore \ 0 \le f_s(\omega) \le \sum_{\alpha=1}^{M} \frac{\rho_{p\alpha}(\omega)}{\bar{A}(\omega)} \tag{C.10}$$

$$0 \le f_s(\omega) \le 1, \tag{C.11}$$

where we used Eq. (C.6) and the fact that the Fermi–Dirac distributions satisfy $0 \le f_\alpha(\omega) \le 1$. The nonequilibrium distribution function $f_s(\omega)$ is thus a linear combination of the Fermi–Dirac distributions of the reservoirs.

C.2 Wiedemann–Franz Law

We explicitly show the derivation of the Wiedemann–Franz law for elastic transport below. The steady-state currents flowing into reservoir p, through a quantum conductor where elastic processes dominate the transport, can be written in a form analogous to the multiterminal Büttiker formula [3]

$$I_p^{(\nu)} = \frac{1}{h} \sum_\alpha \int_{-\infty}^{\infty} d\omega \ (\omega - \mu_p)^\nu \ \mathcal{T}_{p\alpha}(\omega) \left[f_\alpha(\omega) - f_p(\omega) \right], \tag{C.12}$$

where

$$\mathcal{T}_{p\alpha}(\omega) = \mathrm{Tr} \left\{ \Gamma^p(\omega) G^r(\omega) \Gamma^\alpha(\omega) G^a(\omega) \right\} \tag{C.13}$$

is the transmission function for an electron originating in reservoir α to tunnel into reservoir p. Our notation uses $\nu = 0$ to refer to the particle current and $\nu = 1$ to refer to the electronic contribution to the heat current. G^r (G^a) is the retarded (advanced) Green's function. Γ^p and Γ^α are the tunneling-width matrices describing the coupling of the system to the probe and contact α, respectively. In Chap. 4, we wrote down the equations in terms of the *electrical current* I_p instead of the *particle current* $I_p^{(0)}$ above. This affects the definition of the linear response coefficients used in Chap. 4. Naturally, the electrical current is related to the particle current by

$$I_p = -eI_p^{(0)} = -\frac{e}{h} \sum_\alpha \int_{-\infty}^{\infty} d\omega \ \mathcal{T}_{p\alpha}(\omega) \left[f_\alpha(\omega) - f_p(\omega) \right], \tag{C.14}$$

whereas the electronic heat current is simply

$$J_p = I_p^{(1)}. \tag{C.15}$$

Operation within the linear response regime allows one to expand the Fermi functions f_α and f_p to linear order near the equilibrium temperature and chemical potential

$$f_\alpha - f_p = \left(\frac{\partial f}{\partial \mu}\right)\bigg|_{\mu_0, T_0} (\mu_\alpha - \mu_p) + \left(\frac{\partial f}{\partial T}\right)\bigg|_{\mu_0, T_0} (T_\alpha - T_p)$$

$$= \left(-\frac{\partial f}{\partial \omega}\right)\bigg|_{\mu_0, T_0} (-e(V_\alpha - V_p)) + (\omega - \mu_0)\left(-\frac{\partial f}{\partial \omega}\right)\bigg|_{\mu_0, T_0} \frac{(T_\alpha - T_p)}{T_0}.$$
(C.16)

The electrical current

$$I_p = \sum_\alpha \mathcal{L}_{p\alpha}^{(0)}(V_\alpha - V_p) + \mathcal{L}_{p\alpha}^{(1)}\frac{(T_\alpha - T_p)}{T_0},$$
(C.17)

to linear order in the voltage and temperature gradients. Using Eq. (C.16) in Eq. (C.12), we obtain the expressions for the linear response coefficients

$$\mathcal{L}_{p\alpha}^{(0)} = \frac{e^2}{h} \int_{-\infty}^{\infty} d\omega\, \mathcal{T}_{p\alpha}(\omega)\left(-\frac{\partial f}{\partial \omega}\right)\bigg|_{\mu_0, T_0}$$
(C.18)

and

$$\mathcal{L}_{p\alpha}^{(1)} = \frac{-e}{h} \int_{-\infty}^{\infty} d\omega\, (\omega - \mu_0)\, \mathcal{T}_{p\alpha}(\omega)\left(-\frac{\partial f}{\partial \omega}\right)\bigg|_{\mu_0, T_0}.$$
(C.19)

The heat current

$$J_p = \sum_\alpha \mathcal{L}_{p\alpha}^{(1)}(V_\alpha - V_p) + \mathcal{L}_{p\alpha}^{(2)}\frac{(T_\alpha - T_p)}{T_0},$$
(C.20)

where we have taken $\mu_p \approx \mu_0$ in Eq. (C.12) since we are interested in terms up to the linear order. Again, we infer from Eqs. (C.16) and (C.12) that

$$\mathcal{L}_{p\alpha}^{(2)} = \frac{1}{h} \int_{-\infty}^{\infty} d\omega\, (\omega - \mu_0)^2\, \mathcal{T}_{p\alpha}(\omega)\left(-\frac{\partial f}{\partial \omega}\right)\bigg|_{\mu_0, T_0}.$$
(C.21)

The derivative of the Fermi function appears in the expressions for all the linear response coefficients and we may use the Sommerfeld series expansion [3, 4]. We find that

$$\frac{h}{e^2}\mathcal{L}_{p\alpha}^{(0)} = \mathcal{T}_{p\alpha}(\mu_0) + 2\,\Theta(2)(k_B T_0)^2 \mathcal{T}_{p\alpha}^{(2)}(\mu_0)$$
$$+ 2\,\Theta(4)(k_B T_0)^4 \mathcal{T}_{p\alpha}^{(4)}(\mu_0) + \ldots$$
(C.22)

and

$$-\frac{h}{e}\mathcal{L}_{p\alpha}^{(1)} = 4\,\Theta(2)(k_BT_0)^2 T_{p\alpha}^{(1)}(\mu_0) + 8\,\Theta(4)(k_BT_0)^4 T_{p\alpha}^{(3)}(\mu_0)$$
$$+ 12\,\Theta(6)(k_BT_0)^6 T_{p\alpha}^{(5)}(\mu_0) + \dots \qquad (C.23)$$

and

$$h\mathcal{L}_{p\alpha}^{(2)} = 4\,\Theta(2)(k_BT_0)^2 T_{p\alpha}(\mu_0) + 24\,\Theta(4)(k_BT_0)^4 T_{p\alpha}^{(2)}(\mu_0)$$
$$+ 60\,\Theta(6)(k_BT_0)^6 T_{p\alpha}^{(4)}(\mu_0) + \dots, \qquad (C.24)$$

where we use the notation from Ref. [3]: $T_{p\alpha}^{(k)}(\mu_0)$ denotes the k-th derivative of the transmission function $T_{p\alpha}(\omega)$ at $\omega = \mu_0$ and Θ is a numerical factor related to the Riemann zeta function

$$\Theta(k+1) = \left(1 - \frac{1}{2^k}\right)\zeta(k+1). \qquad (C.25)$$

Explicitly:

$$\Theta(2) = \frac{\pi^2}{12}$$

$$\Theta(4) = \left(\frac{7}{8}\right)\frac{\pi^4}{90} \qquad (C.26)$$

$$\Theta(6) = \left(\frac{31}{32}\right)\frac{\pi^6}{945}.$$

The transmission function has appreciable changes on an energy scale determined by the system's Hamiltonian and its couplings to the contacts. We thus define the characteristic energy scale Δ

$$T_{p\alpha}(\mu_0) = \Delta^2 T_{p\alpha}^{(2)}(\mu_0), \qquad (C.27)$$

which is typically much larger than the thermal energy k_BT_0 for most experimental setups.

The following relation connecting $\mathcal{L}_{p\alpha}^{(0)}$ and $\mathcal{L}_{p\alpha}^{(2)}$, from Eqs. (C.22) and (C.24), is the Wiedemann–Franz law:

$$\mathcal{L}_{p\alpha}^{(2)} = \frac{\pi^2 k_B^2 T_0^2}{3e^2}\mathcal{L}_{p\alpha}^{(0)}\left(1 + \frac{8\pi^2}{15}\left(\frac{k_BT_0}{\Delta}\right)^2 + \dots\right). \qquad (C.28)$$

References

1. V. Gasparian, T. Christen, M. Büttiker, Phys. Rev. A **54**, 4022 (1996)
2. S. Datta, *Electronic Transport in Mesoscopic Systems* (Cambridge University Press, Cambridge, 1995)
3. A. Shastry, C.A. Stafford, Phys. Rev. B **92**, 245417 (2015). https://doi.org/10.1103/PhysRevB. 92.245417. http://link.aps.org/doi/10.1103/PhysRevB.92.245417
4. N.W. Ashcroft, N.D. Mermin, *Solid State Physics* (Brooks/Cole - Thomson Learning, Pacific Grove, 1976)

Appendix D
Temperature Calibration and Gating

We discuss here two key topics pertaining to the practical realization of the *scanning tunneling thermometer*.

D.1 Calibration of Temperature

The thermoelectric circuit requires the calibration of the contact temperatures which we describe here. The Pt-heater is fabricated atop an electrically insulating layer above the metal contact α and has a thermal conductivity κ_{hc} with the contact. The temperature of the Pt-heater is inferred from its resistivity. The contact α is heated when an electrical current is passed in the Pt-heater but it also loses heat to the ambient environment which is at the equilibrium temperature $T_{\text{Env}} = T_0$. We denote the thermal conductivity between the contact and the ambient environment by κ_0. The thermal circuit is shown in Fig. D.1.

The heat current flowing into the contact is given by

$$\dot{Q}_{\text{in}} = \kappa_{hc}\left(T_{\text{Heater}} - T_{\text{Contact}}\right) \tag{D.1}$$

whereas the heat current flowing out

$$\dot{Q}_{\text{out}} = \kappa_{0c}\left(T_{\text{Contact}} - T_{\text{Env}}\right). \tag{D.2}$$

In steady state, the rate of heat flow into the contact is equal to the rate of heat lost to the ambient environment and we find

$$T_{\text{Contact}} = \frac{\kappa_{hc}T_{\text{Heater}} + \kappa_{0c}T_{\text{Env}}}{\kappa_{hc} + \kappa_{0c}}. \tag{D.3}$$

© Springer Nature Switzerland AG 2019
A. Shastry, *Theory of Thermodynamic Measurements of Quantum Systems Far from Equilibrium*, Springer Theses, https://doi.org/10.1007/978-3-030-33574-8

When the heater is in good thermal contact $\kappa_{hc} \gg \kappa_{0c}$, we find that

$$T_{\text{Contact}} \approx T_{\text{Heater}}. \tag{D.4}$$

An alternating voltage $V(t) = V_{\max} \cos(\omega t)$ results in a current $I(t) = G_{\text{Pt}} V(t)$ in the heater. The power dissipated via Joule heating is given by

$$P = G_{\text{Pt}} V_{\max}^2 \cos^2(\omega t) = \frac{1}{2} G_{\text{Pt}} V_{\max}^2 \big(1 + \cos(2\omega t) \big), \tag{D.5}$$

which results in 2ω modulations of the heater temperature

$$T_{\text{Heater}} = T_0 + \Delta T_{\max} \big(1 + \cos(2\omega t) \big), \tag{D.6}$$

since the net power dissipated by the heater can be written as

$$P = \kappa (T_{\text{Heater}} - T_0), \text{ where,}$$
$$\kappa = \kappa_{0h} + \frac{\kappa_{hc}\,\kappa_{0c}}{\kappa_{hc} + \kappa_{0c}}, \tag{D.7}$$

as seen from the thermal circuit shown in Fig. D.1.

The temperature of the heater is inferred from the conductance (or resistivity) dependence of the Pt heating element $G_{\text{Pt}}(T)$. The modulation frequency must be chosen so that $\omega \ll 1/\tau$, where τ is the thermal time constant of the metal contact, so that it has enough time to thermally equilibrate. It is of course understood that

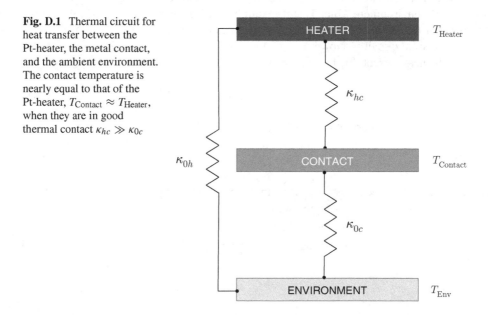

Fig. D.1 Thermal circuit for heat transfer between the Pt-heater, the metal contact, and the ambient environment. The contact temperature is nearly equal to that of the Pt-heater, $T_{\text{Contact}} \approx T_{\text{Heater}}$, when they are in good thermal contact $\kappa_{hc} \gg \kappa_{0c}$

such a frequency allows the heater itself to equilibrate and enter a steady state of heat transfer with the metal contact. The temperature modulations in the metal contact closely follow that of the heater when there is good thermal contact:

$$T_{\text{Contact}}(t) = T_0 + \Delta T_{\max}\big(1 + \cos(2\omega t)\big). \tag{D.8}$$

We have chosen $\Delta T_{\max}$ such that the contact would reach a maximum temperature of $T_0 + 2\Delta T_{\max}$. The calibration fixes $\Delta T_{\max}$ accurately.

We also note that the temperature modulations can be obtained by means other than using a Pt resistor. A graphene flake itself undergoes Joule heating and could therefore be used as a heating element so long as one is able to calibrate its temperature accurately.

D.2 Tunneling Currents

D.2.1 Thermoelectric Circuit

The tunneling current resulting from the heating of the contact is given by

$$I_p = \mathcal{L}_{p\alpha}^{(1)}\frac{(T_\alpha - T_0)}{T_0} \tag{D.9}$$

during the operation of the thermoelectric circuit. Standard lock-in techniques are employed to measure the current amplitude at frequency 2ω. It is easy to see from Eq. (D.8) that the current amplitude

$$I_p\big|_{2\omega} = \mathcal{L}_{p\alpha}^{(1)}\frac{\Delta T_{\max}}{T_0}. \tag{D.10}$$

We show the spatial variation of the tunneling current amplitude in Fig. D.2. The probe is held at a constant height of 3 Å above the plane of the sample. We assume a modest increase in the contact temperature by setting $\Delta T_{\max} = (10\%)\,T_0$ where the equilibrium temperature $T_0 = 4\,\text{K}$. The corresponding contact $\alpha = \{1, 2\}$ is shown by black squares in Fig. D.2 and represent the sites of the sample which are covalently bonded to the metal contact α. $\alpha = 1$ is shown on the left panel and $\alpha = 2$ is shown on the right panel in Fig. D.2. The tunneling current amplitude is as high as 150 pA at some points on the sample and is therefore well within the reach of present experimental resolution. Since we illustrate our numerical results for an experiment performed at liquid He temperatures (4 K), the thermoelectric response is suppressed and gating becomes important. If, for example, T_0 was set to 40 K, we would have a 100-fold increase in the tunneling current amplitude [cf. Eqs. (D.10) and (C.23)] and gating would be less important.

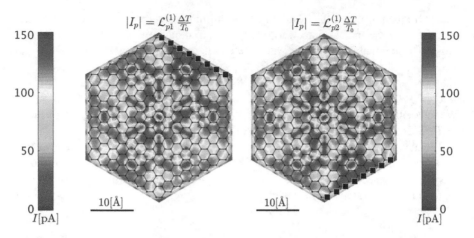

Fig. D.2 Amplitude of the tunneling current in the thermoelectric circuit. The left (right) panel shows the tunneling current amplitude resulting from the heating of the first (second) contact as shown with the black squares in the corresponding panel. The gating potential is set at $\mu_0 = -2.28$ eV with respect to the Dirac point. The amplitude of temperature variations in the contacts [cf. Eq. (D.8)] is taken to be 10% of the equilibrium temperature $\frac{\Delta T_{\max}}{T_0} = 0.1$

Gating

We find that the system has a sufficiently large thermoelectric response at 4 K, i.e., the current amplitude in Eq. (D.10) is experimentally resolvable, when the system is gated appropriately. Indeed, our method works perfectly well for systems which do not have a good thermoelectric response. In such a case, $\mathcal{L}_{p\alpha}^{(1)}$ would have a low value and would result in a current amplitude which is too small to measure. This merely implies that the thermoelectric contribution to the measured temperature is very small—that is, a voltage bias within the linear response regime does not lead to measurable differences in temperatures across the sample. We have chosen the system's gating so that the thermoelectric response is appreciable and there are measurable temperature differences across the sample even in the case of a voltage bias. We find this latter case more interesting.

The thermoelectric coefficient depends on the transmission derivative [cf. Eq. (C.23)] near the equilibrium chemical potential. In Fig. D.3, we show the transmission functions as a function of the chemical potential. The figure shows the transmission spectra into the probe from the two contacts $\alpha = \{1, 2\}$ for one representative point on the sample where the probe is held at a height of 3 Å above the plane of the sample; the transmission spectra would change from point to point on the sample but will roughly resemble the one in Fig. D.3. The contact $\alpha = 1$ is shown in blue (dotted and dashed), whereas $\alpha = 2$ is shown in red. We found that the transmission derivatives are enhanced when the chemical potential is tuned (via the gate voltage) to $\mu_0 = -2.28$ eV and therefore illustrated the thermoelectric circuit for this choice of gating. The resulting temperature measurement is shown in

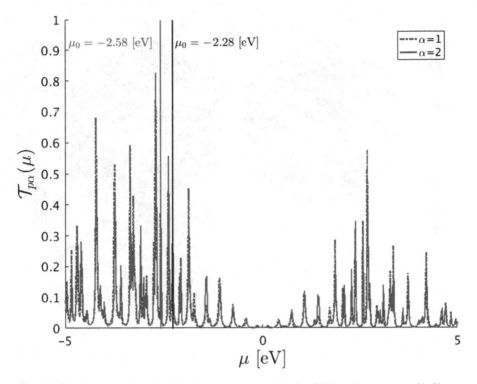

Fig. D.3 Transmission function $\mathcal{T}_{p\alpha}$ from contact α into the STM probe p. $\alpha = \{1, 2\}$ are shown in (dotted-dashed) blue and red, respectively. The conductance circuit measurement is illustrated at a gating potential of $\mu_0 = -2.58\,\text{eV}$ (magenta vertical line). However, we illustrate the thermoelectric circuit at a gating potential of $\mu_0 = -2.28\,\text{eV}$ (black vertical line) since the transmission functions show a large change at that choice of gating, thereby resulting in an enhanced thermoelectric effect

Fig. 4.3 in Chap. 4 for a pure voltage bias. The spatial variations in the transmission derivatives would resemble the pattern shown in Fig. D.2 [cf. Eq. (C.23)].

D.2.2 Conductance Circuit

The tunneling current resulting from the conductance circuit would simply be

$$I_p = \mathcal{L}_{p\alpha}^{(0)}(V_\alpha - V_p). \tag{D.11}$$

We apply an AC voltage $V_\alpha - V_p = V(t) = V_{\text{max}} \cos(\omega t)$ across the contact-probe junction and measure the resulting tunneling current

$$I_p(t) = \mathcal{L}_{p\alpha}^{(0)} V_{\text{max}} \cos(\omega t) \tag{D.12}$$

$$|I_p| = \mathcal{L}_{p1}^{(0)} V_{\max} \qquad\qquad |I_p| = \mathcal{L}_{p2}^{(0)} V_{\max}$$

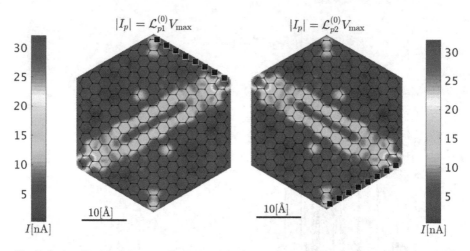

Fig. D.4 Amplitude of the tunneling current in the conductance circuit. The gating potential is set at $\mu_0 = -2.58\,\text{eV}$ with respect to the Dirac point. The left (right) panel shows the tunneling current amplitude resulting from the voltage bias between the first (second) contact and the probe, $V(t) = V_\alpha - V_p = V_{\max}\cos(\omega t)$, as shown with the black squares in the corresponding panel. $V_{\max} = 1\,\text{mV}$

using standard lock-in techniques. The tunneling current amplitude at frequency ω

$$I_p\big|_\omega = \mathcal{L}_{p\alpha}^{(0)} V_{\max} \tag{D.13}$$

is measured across the sample as shown in Fig. D.4. We set the amplitude of voltage modulations $V_{\max} = 1\,\text{mV}$ and a scan of the sample is obtained by maintaining the probe tip at a height of 3 Å above the plane of the sample. The tunneling current amplitude is as high as 30 nA for some regions in the sample. Generally, gating does not play as important a role in the measurement of conductances since we obtain tunneling currents of the order of a few nA for most choices of gating. The corresponding contact $\alpha = \{1, 2\}$ is shown by black squares in Fig. D.4 and represent the sites of the sample which are covalently bonded to the metal contact. $\alpha = 1$ is shown on the left panel and $\alpha = 2$ is shown on the right panel in Fig. D.4. The resulting temperature measurement is shown in Fig. 4.2 in Chap. 4 for a pure temperature bias.

CPSIA information can be obtained
at www.ICGtesting.com
Printed in the USA
LVHW080423131220
674041LV00003B/20

9 783030 335762